Mobile Agents

Advances in Management Information Series

Objectives of the Series

Information and Communications Technologies have experienced considerable advances in the last few years. The task of managing and analysing ever-increasing amounts of data requires the development of more efficient tools to keep pace with this growth.

This series presents advances in the theory and applications of Management Information. It covers an interdisciplinary field, bringing together techniques from applied mathematics, machine learning, pattern recognition, data mining and data warehousing, as well as their applications to intelligence, knowledge management, marketing and social analysis. The majority of these applications are aimed at achieving a better understanding of the behaviour of people and organisations in order to enable decisions to be made in an informed manner. Each volume in the series covers a particular topic in detail.

The volumes cover the following fields:

Information
Information Retrieval
Intelligent Agents
Data Mining
Data Warehouse
Text Mining

Competitive Intelligence
Customer Relationship Management
Information Management
Knowledge Management

J. Rao
Case Western Reserve University
USA

D. Riaño
Universitat Rovira I Virgili
Spain

J. Roddick
Flinders University
Australia

F. Rodrigues
Poly Institute of Porto
Portugal

F. Rossi
DATAMAT
Germany

D. Sitnikov
Kharkov Academy of Culture
Ukraine

R. Turra
CINECA Interuniversity Computing Centre
Italy

D. Van den Poel
Ghent University
Belgium

J. Yoon
Old Dominion University
USA

N. Zhong
Maebashi Institute of Technology
Japan

H.G. Zimmermann
Siemens AG
Germany

Mobile Agents

Principles of Operation and Applications

EDITOR

A. Genco
University of Palermo, Italy

WIT*PRESS* **Southampton, Boston**

Editor

A. Genco
University of Palermo, Italy

Published by

WIT Press
Ashurst Lodge, Ashurst, Southampton, SO40 7AA, UK
Tel: 44 (0) 238 029 3223; Fax: 44 (0) 238 029 2853
E-Mail: witpress@witpress.com
http://www.witpress.com

For USA, Canada and Mexico

WIT Press
25 Bridge Street, Billerica, MA 01821, USA
Tel: 978 667 5841; Fax: 978 667 7582
E-Mail: infousa@witpress.com
http://www.witpress.com

British Library Cataloguing-in-Publication Data

A Catalogue record for this book is available
from the British Library

ISBN: 978-1-84564-060-6
ISSN: 1742-0172

Library of Congress Catalog Card Number: 2007932023

The texts of the papers in this volume were set individually by the authors or under their supervision.

Printed in Great Britain by Cambridge Printing.

Contents

Preface

Multi-agent systems are one of the most effective software design paradigms, and they are considered to be the most recent evolutionary step of object-oriented programming. Agents have several advantages when compared with objects. The most important among them is to be made of active code which is capable of acting autonomously.

Agents can be a suitable choice to exploit the internet reality, since users can operate easily in a less compelling way and also reduce internet connection time. Mobile agents thus make a PC an intelligent entity able to autonomously accomplish boring human tasks, starting from document search up to actual business negotiations. In other words, mobile agents allow the PC-internet system to be more active and autonomous, and leave a human owner to decide if and when his intervention is suitable or required. We can therefore say an agent is an autonomous software entity within the virtual environment in which entities and relationships are devoted to the management and provision of electronic information .

A mobile agent is just an agent endowed with advanced mobility features. In particular we refer to the so-called strong mobility, which gives an agent the ability to accomplish its task migrating from site to site, thus starting somewhere a process to be continued elsewhere.

Mobile agents are a well-known technology by this time, which can be considered a valid alternative to traditional web navigation. Unfortunately, and probably due to the wariness to commit important decisions to virtual entities within a digital environment which does not have countermeasures enough to clash with attacks to data integrity and privacy, mobile agents are not so popular so far. Actually, security is still a problem, not only for mobile agents, but for all software and data within the internet environment.

The book describes the mobile agent principles of operation in detail. It starts from giving some definitions, and illustrates their main features such as mobility, communication, coordination, interoperability, fault tolerance and security. Comparisons of these features between most relevant multi-agent developing platforms are then discussed.

The book ends with a discussion on a mobile agent application field, data mining and information retrieval namely, thus showing how mobile agents can help us to face these field related problems.

The whole work was accomplished with the contribution of students of Operating Systems and Distributed Systems classes provided by Alessandro Genco at the University of Palermo. The final synthesis, revision and management was carried out within the Ph.D. school in Computer Engineering by professors, researchers and doctoral students.

The Editor, 2007

Acknowledgement

This book has been written in collaboration with final-year students in Computer Engineering at the University of Palermo, who where requested to prepare seminars on selected topics of the Operating Systems course, and therefore became co-authors of the chapters.

The Editor is grateful to Stefania Sola and Alessandro Castello for the Italian–English translation of chapters 1-7 and 8 respectively, and to Salvatore Sorce for the text adaptation.

He is also grateful to CNR (National Research Council of Italy) and to MIUR (Ministry of the Instruction, University and Research of Italy) for their financial support.

Intelligent agents

Vittorio Anselmi, Valerio Perna, Vincenzo Spadaro, Maurizio Spataro, Massimo Terranova and Alessandro Genco
DINFO – Dipartimento di Ingegneria Informatica Università degli Studi di Palermo

1 Intelligent agents

Mobile agents intelligence can be traced back to the ability degree the user expects from them. Every agent can be destined to have a limited intelligence based on simple though rigid rules, on some mental notions, on the ability to plan (for example the ability to plan actions that are independent from instructions given by the user) or on learning abilities (for example the ability to learn and adapt to the usual behaviour of the user).

Learning can range from simple events to more complex ones through past experiences (past memories with backtracking capability), so as to be able to elaborate useful models for future occasions.

In latest years the approach used to solve complex problems has shifted from the development of big systems integrated software cooperation to the development of small independent software components able to interact with man, with other software components and with different data sources. Agents can either draw specific or periodic information from precise information sources or execute tasks or services based on acquired information.

2 Classes of agents

Agents' behaviour is determined by their typology. Agents can be classified in various ways: according to their functional features, to the tasks they carry out, to the technology plan, etc.

Intelligent agents classification is not exclusive, since some of the functional features used for the classification are not mutually exclusive.

It is useful for our purposes to have a look at some intelligent agents classifications proposed by some artificial intelligence experts.

2.1 Nwana classification

Nwana [1] singles out a typology containing four classes of agents differentiated according to their ability to cooperate, learn and act autonomously.

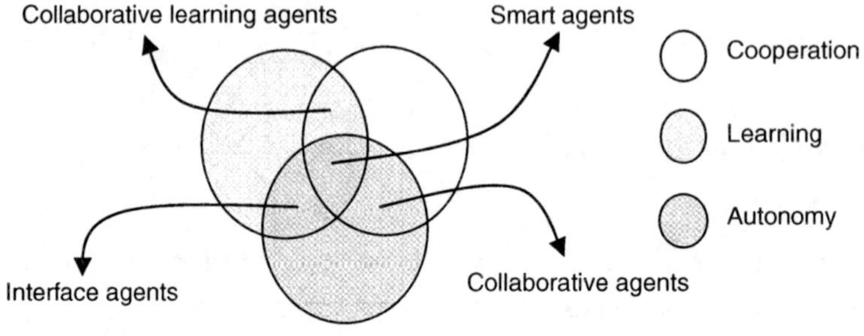

Figure 1: Nwana classification of agents.

It is possible to provide a graphic representation (fig. 1) showing how four different classes of agents can spring from the combination of the above-mentioned three functional features.

It is important to note that such classification is not rigid, since the agent's belonging to a class does not deny it the lacking ability. That is why for collaborative agents, for example, favourite aspects are those of cooperation and autonomy rather than the learning one. It does not mean, however, that a collaborative agent is not able to learn. Similarly, in carrying out interface agents favoured aspects are those of autonomy and learning rather than the cooperation one. Generally speaking, Nwana claims that an ideal agent should equally have all three abilities.

2.2 Davis classification

D.N. Davis [2] describes three types of agents' behaviour, sorted into increasing level of intelligence: reflective, reactive and meditative.

Whereas reflective and reactive agents do not have explicit motivational states, meditative agents can even reason on their objectives. Reflective and reactive agents are usually included in the class of reactive agents, while meditative agents form the so-called class of deliberative agents. From the union of these two classes a third one can be formed, which contains the characteristics of both classes. The agents belonging to the latter are called hybrid agents.

Hereafter is a graphic representation (fig. 2) of this classification.

2.3 Reactive agents

Reactive agents, otherwise called stimulus–response agents in the field of Artificial Intelligence, give a valid solution to the exigency of instant reactions by an agent as an answer to the changes of a dynamic environment, and to the presence of nodes with limited computational power and low transmission band.

Such agents are not able to plan, thus proving to be limited in the choice of their actions, since these are selected only on the basis of present perceptions of

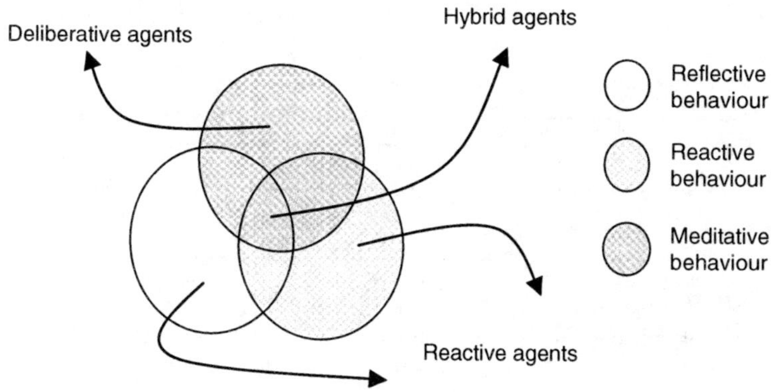

Figure 2: Davis' classification of agents.

environment. Moreover, they do not normally carry out a scheduling on actions and do not know the modifications made on environment by their actions, nor the final goal they are trying to pursue.

It should however be noted that, according to this reasoning, the behaviour of several animals, a lot of insects, for example can be considered as a form of intelligence. A familiar example to everybody is the behaviour of some night butterflies attracted by the light of a common incandescent lamp (the objective) that, though keeping on bumping against it, as soon as they recover their strength, begin again to make for the light source, sometimes being killed by the repeated burns.

It seems, therefore, that the attribute "intelligent" does not suit completely to this class.

Brooks [3] states that this kind of agents' intelligence is not to be found in the agent itself, but in the society of agents and in the environment in which they work. Let us consider, for instance, once again the animal world, particularly ants. On the one hand, each one does not but repeat endlessly the same tasks it has been prepared for (food collection, nest maintenance, territory defence); on the other hand, if we look at them as a whole, we cannot but attribute the whole system an intelligent behaviour.

On the face of it, one could also question the usefulness of giving an intelligent agent such a behaviour. Actually, in the light of some solutions, an approach taking into account the use of an agent aware of external environment, of the consequences of its actions, of the goal to be pursued – all requirements lacking to stimulus–response agents – is undoubtedly preferable. It is to be noted, however, that those disadvantages can be compensated by three advantages a stimulus–response agent possesses:

- Remarkable production simplicity
- Extremely low time of response
- Fault tolerance

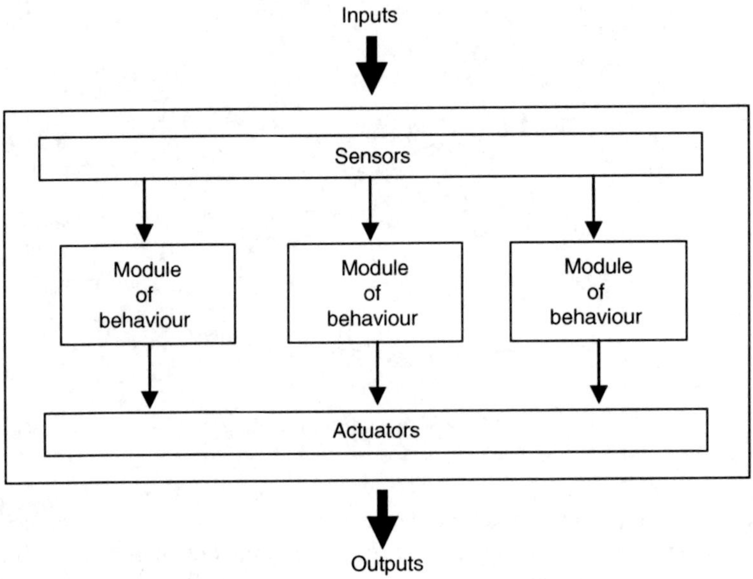

Figure 3: General architecture of a reactive agent.

Production simplicity is due to the absence of learning and reasoning elements that normally need a particularly complex development. As regards response time, it should be noted that a reactive agent does not have to elaborate the data received by its own sensory inputs, so the only factor determining response time is the speed of the agent in receiving such input.

In the light of what we have just seen, we must acknowledge that there can be cases in which this kind of solution is preferable, when particularly simple tasks or real-time execution is required.

In order to give a full treatment of the topic, fig. 3 reproduces the general architecture of a reactive agent [4].

The figure clearly shows that the architecture of a reactive agent takes into account the presence of different behavioural modules, each of them carrying out a specific task without any connection to the other ones.

A robot, for instance, can have a module enabling it to walk getting over obstacles, another one to decide where to go and others to carry out any other behaviour. It can be therefore noted that the plan of a reactive agent is already inherent in the agent's nature.

It can be easily seen then that no central planning and reasoning entity is present. All the single behavioural modules work in parallel and elaborate their own input in order to produce their own output. This kind of architecture is very tolerant of faults; any wrong functioning of a module does not jeopardize the general functioning of the agent, nor that of the other modules.

2.4 Deliberative agents

Deliberative agents are a valid alternative to reactive agents. They have a symbolic model of their physical environment logic connections. These agents create and follow some plans to achieve a goal, by processing the inputs and restoring an output totally. There are no modules working independently, but independent modules working together.

Deliberative agents are usually also called BDI Agents (Belief–Desire–Intention Agents), since the "Belief, Desire, Intention" modules are the main components of this kind of agent's internal conditions [5]. Belief, Desire and Intention represent the information, motivation and deliberation states of the agent. These mental dispositions determine the agent's behaviour and are of basic importance to carry out optimal performances when the decision is subject to resources constraints [6, 7].

Before going on, it is necessary to understand what the three essential components: Belief, Desire and Intention really are.

Since the actions the agent must perform to achieve a goal do not depend on its inner state, but on the environment context, it is necessary that the agent itself has got information about the real state of environment. Since such information cannot be gathered with a single act of perception, it is necessary for the agent to have a state component representing this kind of information, to be suitably updated after each perception. Such informative component is called Belief. It can be thought of as a variable, a database or a data structure suitably projected for each case. It is furthermore necessary for the agent to have information about the goals to be pursued, or, generally speaking, to know the priorities associated with the various current goals.

These goals can be possibly generated immediately or through a function, in which case a representation of the state is not necessary. Contrariwise, beliefs cannot be generated through functions. This component is called Desire. It can be thought of as the agent's motivational state.

The choice of the actions to be undertaken needs from the agent a certain computation time within which the environment can change. This would nullify the agent efforts, since the action chosen could not bring anymore to the achievement of the current goal. One possible solution could be that of controlling the environment state validity step by step, or that of carrying the action out disregarding any possible environmental change. In any case, assuming that significant changes can be determined almost immediately, it is possible to limit reconsideration frequency and thus reach a balance between too much and not enough consideration [7]. It is necessary for this purpose to include a state component of the agent in order to represent the current course of actions. Such additional state component is called Intention. Ultimately, intentions are the agent deliberative component.

Starting from Davis's premises, Brenner [8] proposes a general architecture for BDI agents, producing a symbolic model of logical connections in its physical environment. The architecture proposed is shown in fig. 4.

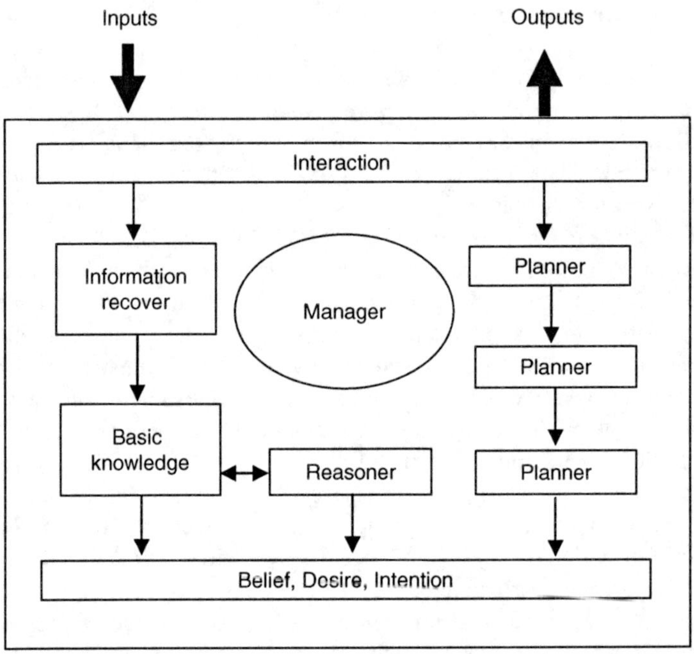

Figure 4: General architecture of a deliberative agent.

It is worth noting the *Reasoner* block inside the model, allowing the agent to elaborate the basic knowledge and formulate desires, goals and intentions. The planner, on the contrary, has the task of gathering intentions and arranging them in plans later scheduled and executed.

This is just one possible model among the ones that can be created. We could think of simpler ones or more complex ones according to our needs.

A deliberative agent is usually characterized by the following basic components:

- A symbolic model of the environment where it has to work.
- A symbolic list of possible actions.
- A planning algorithm having an input representation of environment, goal and the list of possible actions, as well as an output sequence of actions that the agent can execute to reach its goal.

Therefore, an intelligent agent selects a goal, creates a list of executable actions, executes them and achieves the already planned goal.

On the one hand, deliberative agents offer considerable potentialities; on the other, they have great disadvantages. Models like this are still very complex to be adapted to sudden environmental changes. This is certainly one of the reasons why deliberative agents are not yet used in dynamic environments.

In order to take advantage of the positive aspects of both classes of agents, it is possible to envisage a third class taking into account the three behaviours proposed by Davis. Such class is called hybrid (hybrid agents).

3 Agents properties

In order to give a definition of agent referable to the concepts of cooperation among different community groups for an intelligent management of information, it is necessary to list the following basic properties:

Autonomy: the ability of the agent to operate without necessarily being imposed any external instruction. The agent must be able to do its own choices and take its own decisions without the intervention of any other super-ordered entity (e.g. the user).

Communication skills: the ability to establish communicative relations with the surrounding environment. Interaction with the user and with similar agents is of basic importance.

Reactivity: the environment surrounding the agent is typically subject to sudden changes continually challenging its capacity of answering the external stimuli so as to consequently adapt its activity.

Mental notions: during its life cycle, the agent must cope with various situations from which it learns. Thus, through the memories of such experiences and through the interaction with the environment and with the other agents, it is able to acquire its own knowledge.

Persistence: a mobile agent life is usually longer than the basic tasks length it achieves. Its existence goes on with an inner state, in order to be able to achieve future interactions. There is no guarantee though that this permanence capacity will be indefinitely maintained. A mobile agent is usually created to satisfy one of the following two persistence criteria:

- To finish after having done the different interactions required by the basic task it had been assigned.
- To finish after having exhausted the assigned internal resources that can be reproduced or used up.

Vitality: during its life cycle, the agent must cope with anomalous situations that create a state of instability that could damage or irremediably jeopardize its persistence. An agent equipped with high vitality is able to solve the most adverse situations it meets.

Mobility: a mobile agent already has the inner attitude to vary its communication partners. It is able to interface itself with both the user and other similar agents without distinction.

Social ability: the ability to communicate with other agents, even cooperating to pursue goals through information and knowledge exchange. This characteristic can be developed through the Agent Communication Language (ACL).

Pro-activity: a mobile agent acts to pursue its arranged goals showing, thus, opportunist behaviour. Pro-activity expresses the ability to generate events in the surrounding environment, to start new interactions with other agents and to coordinate the various agents' activities stimulating them to give rise to certain responses.

Truthfulness: the agents system takes into account exchange of information that is not under direct control of super-ordered inspection entities, so they are expected to respect truthfulness.

Benevolence: a mobile agent endowed with the intelligence characteristics above mentioned must not perform acts contrary to the user's will.

Mobile agents systems can be compared to distributed objects systems, as both involve entities endowed with their own inner state. There are though some differences that make the use of mobile agents preferable: first, their autonomy allows them to keep under complete control their actions, while objects system does not allow to achieve the same control level. ACLs are moreover independent from applications.

Agents supply us with metaphors useful for the description of artificial systems, as the following:

- Open systems: they vary in a dynamic way since they are based on heterogeneous components that appear, disappear and change their behaviours.
- Complex systems: mobile agents give an analysis and synthesis method.
- Distributed data, control, or resource systems: solutions are given for their own implementation.
- Legacy systems: solutions are given to foster interoperability among pre-existing older software.

4 Complexity and coherence

Intelligent agents need to be considerably complex in order to perform their tasks. The technique used to treat complexity is abstraction. J.R. Rose and M.N. Huhns [9] believe that the abstraction kind and level suitable to treat intelligent agents can be found in philosophy.

In order to give reliability to the system, however, single agents need basic principles that, in turn, may ensure the entire system reliability.

4.1 Global coherence

An agent-based approach is, by its very nature, distributed and autonomous; but, when communication channels are noisy or have low bandwidth, agents will have to take decisions locally, with the hope of global coherence.

We can trust agents working locally if they use ethical principles we understand and share.

In order to equip an agent with ethical principles, developers need an architecture supporting explicit goals, principles and abilities (for example how to negotiate), as well as laws and means to impose a sanction on or punish possible transgressors.

Coherence is described as the absence of useless efforts and progress towards the designed goals. Within an architecture of agents supporting both trust and coherence, the lowest level enables an agent to behave in a reactive way, i.e. to react to immediate events. Intermediate levels deal with agents' interaction, while the highest levels enable the agent to take into consideration its behaviour's long-term effects on the rest of its society. Agents are typically planned starting from the base of this architecture, with growing abstract reasoning ability as they go up the scale. The awareness of the presence of other agents and of their role in a society, implicit in the level of social commitment and in higher ones, can give the agent the possibility of behaving coherently.

5 Ethical abstractions

Ethics is a branch of philosophy that deals with moral behaviour codes and principles [9]. Many ethical theories distinguish between right and good: right is what is right in itself. Good is what is good or valuable to somebody or to some aim. The so-called deontological theories emphasize the concept of "right rather than good". They oppose the idea that the end justifies the means.

These theories distinguish between intentional effects and unexpected consequences. That is to say, an action is not wrong unless the agent's intention is to explicitly harm through it. This legitimates inactivity, even when inactivity has predictable though unwelcome consequences. Teleological theories, on the contrary, chose good rather than right: something is right if it maximizes good; in that case the end can justify the means. In teleological theories, honesty of action is based on actions ability to satisfy different goals, not on their inner goodness.

The choice of actions can be based on comparison or on preference. Selfishness and utilitarianism are ethical theories parallel to each other: on the one hand, utilitarians assert that action should maximize the universal welfare of all agents; egoists assert, on the other hand, that action should maximize their own interests.

Both are teleological theories, since both assert that the right thing to do is to produce a certain welfare. What agents need in order to choose their actions are not only universal principles or consequences, but also a certain consideration of their promises and duties. Their *prima facie* duties are to keep promises, help others, repay courtesies and so on, as long as these duties are performed without violating any principle and as long as there is not a more important duty to be performed.

Ethical theories are justificatory rather than deliberative theories. An agent can decide which basic value system use according to the approach chosen.

Deontological theories are more cogent and ignore practical considerations, but they must be thought of as incomplete bonds, so that the agent can chose one right action among others.

Teleological theories are broader and include practical considerations, but they leave the agent less free to choose the best alternative.

All ethical approaches are single agent oriented and implicitly codify other agents.

6 Intelligent communication languages

Client–server architecture classes (fig. 5) can communicate thanks to a standard protocol established by the International Standard Organization/Open System Interconnection (ISO/OSI).

Today, we do not have an analogous standard yet as far as the basic feature of intelligent agents, i.e. cooperation among mobile agents, is concerned. Studies about ACL are of basic importance for future development of research.

The agents' programmer has at his disposal two alternatives:

- To create a dedicated protocol or language allowing agents to communicate mutually; thus, he isolates them and prevents any outside agent from communicating with them.
- To use a communication language created by other researchers, thus widening the class of agents with which his system can communicate.

The first step for the establishment of a standard ACL is, first of all, to decide which basic requirements it will have, secondly to give the agents a common syntax, semantics and pragmatics. The more the following basic requirements for the creation of a possible ACL standard will be carried out, the more reliable intelligent communication will be:

Form: a good ACL should be syntactically simple, concise and easily readable by programmers, easy to analyse and create.

Content: a good ACL should possess a well-defined set of extendable primitives, in order to not only ensure the use of language in a wide range of systems, but also assure its use among different applications asking for or offering services.

Semantics: semantics must define the effect of each single operation, expressed in terms of pre-conditions and post-conditions. Unambiguousness of the operations thus expressed is also needed.

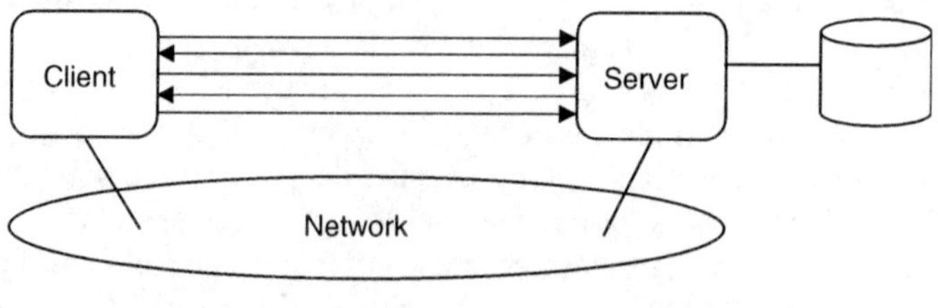

Figure 5: Client–server architecture classes.

Implementation: implementation should be efficient in terms of both speed and utilization. It should well adapt to existing technologies, and possess an easily usable interface. It should, finally, be adaptable to any kind of language, be it an object one like C++, Smalltalk, Eiffel, Java, or be it a procedural one like C and Lisp.

Networking: a good ACL should well adapt to modern network technologies, supporting all kinds of basic connection. Furthermore, it should contain a high number of communication primitives to be used in different languages and protocols.

Reliability: it is a basic requirement that should be met giving different safety options, such as to ensure private exchange of information between two agents; to possess a method ensuring authenticity of the agent with whom one is communicating; and to be fault tolerant.

7 Mobile agents training

The methods allowing agents to acquire a certain behavioural intelligence can be classified as follows:

User looking: the agent improves its knowledge observing for a certain period the user even without his knowing it; it keeps trace of choices and consequently changes its profile.

User indirect feedback: the agent suggests results to the user, and takes note if he ignores the suggestion doing sometimes the contrary.

User direct feedback: the agent improves its knowledge by asking the user explicit explanations about his choices.

Learning by example: the user can give the agent a range of examples as a basis on which it can work. Two different approaches are possible:

Static knowledge: examples given in advance to the agent for its training are very often hypothetical special cases.

Dynamic knowledge: examples in question are taken from real situations and shown to the agent by the user as they occur.

Agent asking: if the agent does not have any specific knowledge about a subject, he asks other agents to give it useful information to solve the problem.

8 Agents systems implementation

We shall hereafter use the diagrams developed by J.M. Vidal, P.A. Buhler and M.N. Huhns [10] in order to give a model based on object-oriented programming. In particular, these diagrams have been created with reference to Unified Modelling Language (UML) [11].

Using conventional object-oriented planning analyses and techniques, it will be possible to stress the agent's being more than a simple object. Following the above-mentioned authors' path, let us take into consideration the following functional characteristics for the implementation of an intelligent agent:

- single identity
- pro-activity
- persistence
- autonomy
- social ability.

As far as single identity is concerned, the agent inherits its own by simply being an object. In order to have pro-activity, an agent must be an object possessing an inner cycle of events similar to that possessed by an object extending the Java thread class. Here is a possible pseudo-coding for a typical events cycle, resulting from environment perception:

```
Environment e;
RuleSet r;
while (true)
       {
       state =senseEnvironment(e);
       a =chooseAction(state,r);
       e.applyAction(a);
       }
```

The printout shows an infinite cycle that gives persistence to the agent. Persistence also makes possible for an agent to learn from others, as well as mutual modelling. To do that, agents must be able to distinguish an agent from the other, hence the necessity of a single identity.

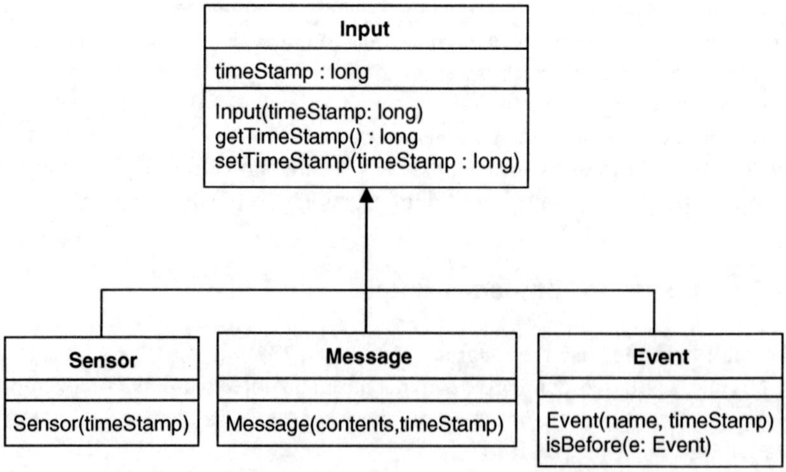

Figure 6: Three possible examples of agent input.

In order to give autonomy to an agent, it is enough to declare all its methods as private. Thus, the agent alone can call for its methods, under its full control, while no other agent can force it to do something it does not want to, by retrieving one of its methods. Giving an agent the ability to communicate with other agents, we obtain social ability, which, as we have already seen, allows the agents to coordinate their actions with those of other agents, and so to cooperate with them [12]. In order to obtain social ability, we can generalize the agent's input class, as shown in UML diagram in fig. 6. UML diagrams can valuably help us understand and develop software agents. It must be said that UML diagrams do not claim comprehensiveness in functional treatment, but provide a general structure for the implementation of agents architectures [10].

As can be seen, in this case three possible input classes have been defined: Sensor, Message and Event. The objects in the Sensor class notably locate every behavioural input detected by the agent sensors, while those in the Event class are used by the agent itself as a reminder.

An agent, for instance, that intends to wait no more than 5 min to receive an answer, could set up an event to be launched after 5 min. If the answer arrives before the event, the agent can disable it. If, instead, it receives the event, then the agent knows that it has not received the expected answer in time and can behave consequently.

8.1 Reactive agents

We have already seen the structure of a reactive agent. It is the simplest to achieve, since the agent does not keep information about its environment, but simply instantly reacts to environmental changes. Fig. 7 shows the class diagram of such agents.

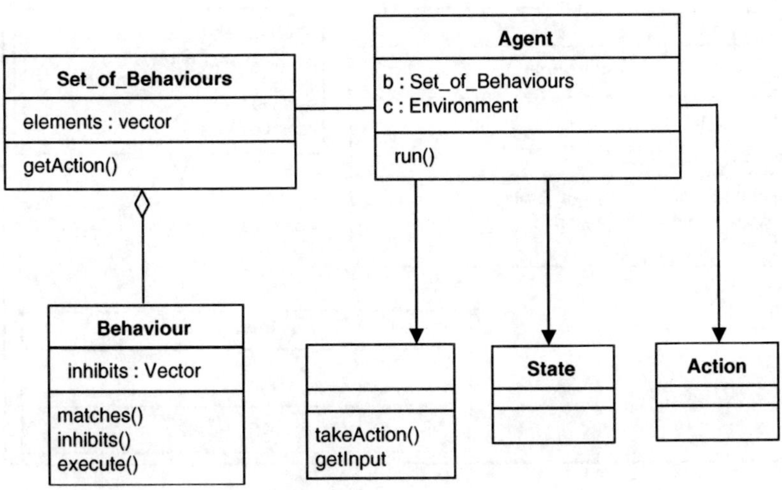

Figure 7: Class diagram of a reactive agent.

8.2 BDI agents

Some implementations of BDI agents define a new programming language and an *ad hoc* interpreter that can interrupt the program at any moment, save the state and execute some other necessary plans.

The solution proposed in the class diagram in fig. 8 does not work in the same way. It makes use of an intentionally multitasking method where a thread constantly controls environment, in order to ensure current intention applicability.

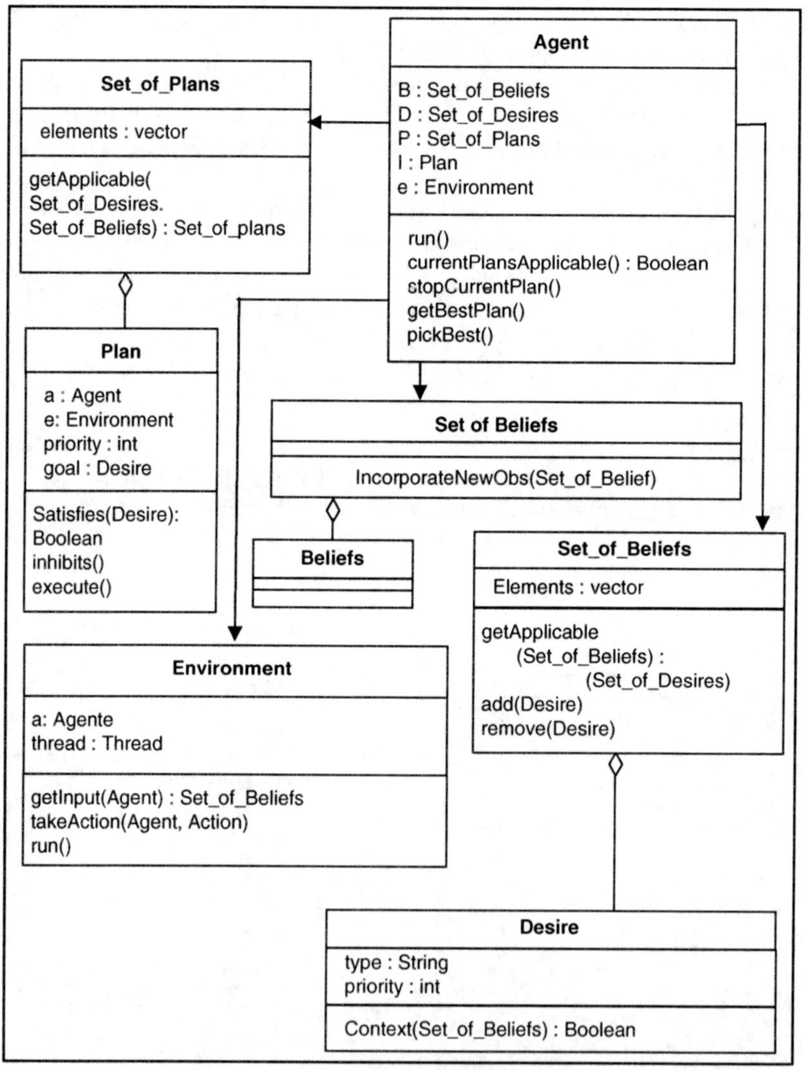

Figure 8: Class diagram of a BDI agent.

If it does not happen, the environment thread will communicate the intention to stop. In order to stop, it will invoke the stopCurrentPlan() method, that, in its turn, will call for the stopExecuting() method. Thus, the plan is responsible for its own interruption and its own erasure.

Giving each plan such ability, the possibility of a deadlock – resulting from the fact that a plan can still have hidden resources after being stopped – is excluded.

9 Behaviours and actions management

Many of the most common agents architectures, including the ones we have analysed before, contain a set of behaviours and a method for scheduling. Behaviour can be distinguished from action because an action is an atomic event, while behaviour can take place in a longer period of time. In multi-agents systems (MAS) it is also possible to distinguish between physical behaviours that generate actions and conversations among agents. We can consider behaviours and conversations as classes inherited from an abstract class of activity. We can thus outline an Activity Manager that is responsible for the scheduling of activities. This activity manager model is suitable for the implementation of many of the most common agents architectures, keeping at the same time the characteristics of encapsulation and modularity required by a good object-oriented programming.

Going into details, activity is an abstract class that defines the interface to be implemented by all behaviours and conversations. Behaviour class can implement the auxiliary methods necessary in a particular field, for example for the triangulation of an agent position; conversation class can implement a finite state machine that can be used for particular conversations; thus, for example entering the right state and adding functions to manage transactions, an agent can define a negotiation protocol as a class inherited by the conversation one.

The details of how this can be achieved depend on how the conversation class implements a finite state machine, which changes according to real-time execution demands.

Defining every activity as an independent object and implementing a different manager for every activity implies considerable advantages. The most important one is separation between knowledge domain and its control, a feature mainly emphasized by blackboard systems.

Activities contain the whole knowledge concerning the particular world where the agent lives, while activity manager embodies knowledge concerning deadlines and other scheduling constraints the agent must deal with. Thus, implementing each activity as a different class, the programmer will have to separate the agent's abilities in encapsulated objects that can be used again by other activities.

Activities hierarchy forces all activities to implement a minimum interface that will be easily used again thanks to inheritability.

9.1 DACS and IMA

Decentralized Autonomous Cooperative Systems (DACS) are sets of independent computational resources that cooperate to provide the basic functionalities in order to integrate the high-standard services on which a distributed system is based. The system is defined as decentralized because there is not a dedicated entity able to provide such functionality. Autonomy derives from the fact that the system represents an open computational environment where hosts can take part and autonomous decisions are taken concerning the actions necessary to achieve the goals required within their domain. As to coordination, single hosts must be enabled to gather information about the state of the system and to communicate actions and decisions to the other ones.

In a global context, precision and uncertainty of decisional information are strictly linked to coordination and individual decisions. Generally, there are some limitations concerning the availability of resources that can be extended to the acquisition and representation of a global view. Generally speaking, we should expect that single entities inside the decentralized system show some typologies of rational behaviour.

Rational behaviour, if informally defined, can be identified with the ability of choosing the best one among available sets. On terms of functioning usefulness, an entity will choose an action in order to maximize usefulness itself. The meaning of usefulness or preference within a DACS has not been defined and is not obvious at all. We have in fact to acknowledge that the subsystems forming DACS are themselves DACS with their preference concepts. It goes without saying that DACS can show rational behaviour at different levels and are not necessarily consistent.

At the user's level, he can expect the system to work to optimize current application.

At the system's level, we can expect the system to optimize its performances in all current applications.

Furthermore, DACS must be able to optimize system activities, providing basic functionalities and searching for the best result. A mutual exchange among single-level preferences will be necessary in order to render multilevel cooperative system. An example of mutual exchange is the need to delay or reduce computation application in order to optimize the access at communication level. In concrete terms, DACS entities must classify and order every possible action based on information about the global state of the system.

Knowledge acquisition (KA) and knowledge representation (KR), together with a system state global view computation, collide with DACS size and complexity.

New KA and KR are necessary to select information from knowledge resources so as to generate an adequately precise system vision. Some examples of problems concerning KA and KR mechanisms are resource monitoring and performances, resource discovery and systems loading management.

In the last decades research into the field of distributed systems has reached good results in the study of innovative mechanisms allowing hosts to cooperate in order to create a single shared computation platform; nevertheless, algorithms developed along these years are not easily adaptable to dynamic exchange facilities, and furthermore they are not adaptable to specific cooperation environments.

Current research pays attention to mobile code in the shape of intelligent mobile agents (IMA) and active networks, offering new possibilities for the development of solutions suitable for a multitude of DACS.

IMA well describe the concept of mobile computation or mobile code. Mobile code is orthogonal to the well-known remote procedure call (RPC), where data are looked for by the program rather than being transferred to the running program.

Contrasting examples of programs have been proposed; they focus on experimental project, implementation and IMA analysis as integral to DACS. Such agents implement automatic working systems and work without user's explicit control. It goes without saying that autonomy is an important aspect of these agents. Instead of working in the user's name, IMA for DACS are considered as system agents working in the name of or to help the system itself. Generally speaking, IMA approach is considered suitable for problems that require a system able to autonomously get around different tasks within a dynamic and unpredictable environment.

Within networking, mobile agents can be seen as catalysts for intelligent services supply inside the network, very similarly to the concept of router inside active networks. IMA are the bases for the development of new mechanisms allowing the network to maintain quality of service (QOS) demands.

In big decentralized systems cooperation among agents is used for distributed allocations and resources sharing. IMA manage independently single resources and cooperate with each other for a global sharing of all the resources, among which the system utilization increases.

9.2 IMA in multi-agents systems

Multi-agents systems area is grounded in the field of distributed artificial intelligence. The first results of research have shown solution to distributed problems and modules coordination in computational components in order to find efficient solutions. In latest years, single IMA and systems composed of several IMA have shown a certain degree of anthropomorphism. That is why IMA have been seen as entities endowed with implicit intellect, thus allowing information technology researchers to study MASs' sociological aspects.

Talking about a DACS, a more complex approach could be used thanks to the analysis of applications complexity and to the way agents carry out their inner properties within the solution space and the environment they work in. It is also important to dwell upon agents' performances in a specific environment.

Though agents working in a DACS can act autonomously, they coordinate their efforts through information sharing and subsequent synchronization implicit in their actions.

Actually, agents' communication to exchange information implies an overload. That suggests the existence of a critical number of agents for a certain environment. Autonomous control of agents' population as function of a system's topology and dynamism represents one of the subjects research is debating on lately in IMA field.

9.3 IMA biological paradigm (cyber entities)

A wide range of biological systems will surpass debates on future applications of networking such as scalability, adaptability and survival/availability. Consequently, future network applications will benefit from this biological view for the development of new principles and mechanisms.

The most interesting applications concern Bio-Networking Architecture (BNA) that can be seen as a set of autonomous mobile agents called cyber entities (CE) used to implement an application.

Cyber entities are nothing but autonomous mobile agents used to implement network applications, while Bio-Net provide execution environments and support services for cyber entities.

A typical BNA paradigm is based on some principles and mechanisms hereafter listed:

- Emergence
- Autonomous actions based on local information and interaction
- Birth and death as expected events
- Energy and adaptation
- Natural selection and evolution

9.4 A cyber-entity paradigm

A CE is an autonomous mobile agent; several CEs are used to create applications that can be thought of as a combination of characteristics, such as attributes, behaviours and body.

Attributes. They are variables that describe CE. Some examples can be owner's name (ownerName), unique identity (uniqueID), date of birth (timeBorn) and energy level (energyLevel).

Behaviours. They are the executable code that implements CE's functionality and the autonomous control of its actions. Among various behaviours we can mention the following:

- the action of reception of an event
- the action of fixing time
- the action of wasting energy
- the action of migration
- the action of reproduction

- the action of receiving messages
- the action of support
- the action of death

Body contains data connected to CE's support action. For example if CE's support action is to transmit web pages, its body will contain web pages.

When a CE reproduces itself asexually, the son's body will be the exact copy of the parent's one. When two CEs reproduce themselves sexually, the son's body will contain the exact copy of one parent's body or of both.

References

[1] Nwana, H.S., Software agents: An overview, *Knowledge Engineering Review*, **11(3)**, pp. 1–40, 1996.

[2] Davis, D.N., Reactive and motivational agents: Towards a collective minder. *Lecture Notes in Artificial Intelligence 1193, Intelligent Agents III*, eds. J.P. Müller, M.J. Wooldridge & N.R. Jennings, *Proc. of ECAI'96 Workshop*, ISBN 3-540-62507-0, Springer, 1997.

[3] Brooks, R.A., Intelligence without representation. *Artificial Intelligence*, **47**, pp. 139–159, 1991.

[4] Palensky, P., The Convergence of Intelligent Software Agents and Field Area Networks, IEEE 0-7803-5670-5/99, 1999.

[5] Rao, A.S. & Georgeff, M.P., BDI agents from theory to practice. *Proc. of the 1st Int. Conf. on Multi-Agent Systems* (*ICMAS*), San Francisco, 1995.

[6] Bratman, M.F., *Intentions, Plans and Practical Reason*, Harvard University Press: Cambridge, MA, 1987.

[7] Kinny, D. & Georgeff, M., Commitment and effectiveness of situated agent. *Proc. of the 12th Int. Joint Conf. on AI, (IJCAI*), Sydney, pp. 82–88, 1991.

[8] Brenner, W., Zarnekow, R. & Wittig, H., Intelligente Software Agenten, ISBN 3-540-63431-2, Springer, 1998.

[9] Rose, J.R. & Huhns, M.N., Philosophical Agents, IEEE Internet Computing, 2001.

[10] Vidal, J.M., Buhler, P.A. & Huhns, M.N., Inside an Agent, IEEE Internet Computing, 2001.

[11] Fowler, M., *UML Distilled, 2nd Edition: A Brief Guide to the Standard Object Modelling Language*, Addison Wesley Longman: Reading, MA, 2000.

[12] Weiss, G., *Multiagent Systems*, MIT Press: Cambridge, MA, 1999.

Mobility

Marco Martorana, Marco Sotgia and Alessandro Genco
DINFO – Dipartimento di Ingegneria Informatica Università degli Studi di Palermo

1 Strong and weak migration

The state of a mobile agent can be divided into two parts (fig. 1): *runtime state* and *data state*. Runtime state contains all information for the control of a mobile agent and is mainly composed of the program counter and the stack. The running agent data state contains information such as the local variables and the basic attributes.

Two types of agent migration can be distinguished: *Strong Migration* and *Weak Migration* [1], according to whether it transfers the runtime state or not (fig. 2).

Strong migration is also called *Transparent Migration*: when the agent asks for migration, both the runtime state and the data state need to be saved and transferred to the destination host together with the agent's application code. When it reaches its destination, the agent can re-obtain the state it had before its migration and can be launched again from exactly the same code position it had before the migration request.

Weak migration is also called *non-Transparent Migration*: when the agent asks for migration, the data state only is saved and transferred to the destination host together with the agent code. When it reaches its destination, the agent must not be launched from the code position it had before but from the main application function.

Both methods require the state saving and later the restoring of the state previously saved, before the agent itself begins again its execution.

In order to correctly carry out strong migration, it is necessary to know the methods contained in the *method call stack* and of the *program counter*. The *call stack* contains all the running methods and their call sequence. Object data include the *object members* and the local variables of all methods lying on the call stack. One of the major problems of Java for the achievement of a transparent migration is that its classes are first interpreted and later run by the Virtual Machine (VM). Therefore, only a limited access to the inner information of the program, such as the *program counter*, the *local stack frames* and the running open resources of the threads, is allowed. Keeping this problem in mind, we show a classification of the different aspects concerning migration in fig. 3.

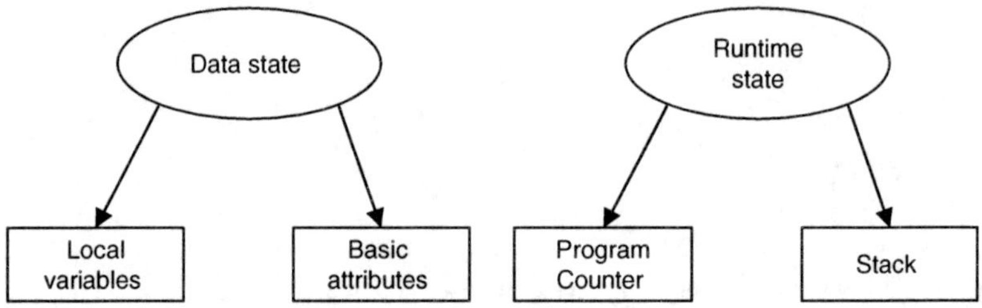

Figure 1: Data state and runtime state.

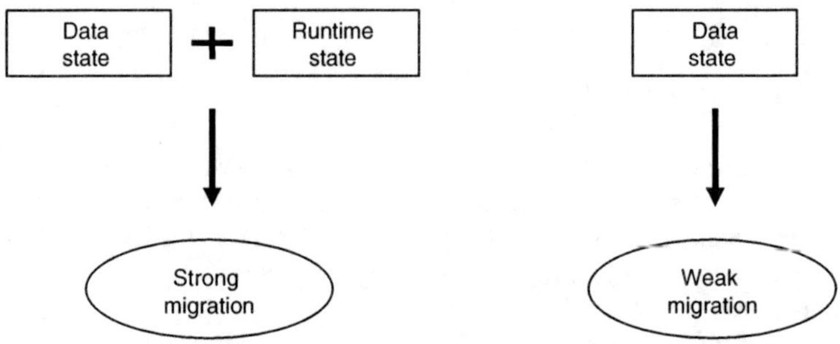

Figure 2: Strong and weak migration.

At the highest level, classification consists of two orthogonal aspects: *code migration* and *state migration*. The former refers to the transfer of code, the latter to the transfer of the agent's state. This is how strong migration is usually described; state migration consists of much more aspects. At the second level, we can find *execution migration* and *data migration*, which means that the state of an agent generally consists of the execution current point and the agent's current data. Both execution and data migration consist of different parts. Execution migration is composed of *program counter migration* and *thread migration. Stack migration, member migration* and *resource migration* constitute data migration. Fig. 3 also shows two other types of weak migration that can be considered an alternative to program counter migration: *initialization migration* and *method migration.*

1.1 Code migration

When this type of migration occurs, the entire agent code needs to be transferred from the source host to the destination one. Since agents do not generally consist of a single class but keep reference to other objects, it goes without saying that these parts of code should be transmitted as well. Nevertheless, code migration should, if needed, transfer the reference objects code *tout court.* This means that the code of Java classes and an *agent framework* classes might also be triggered as soon as the agent reaches its destination host.

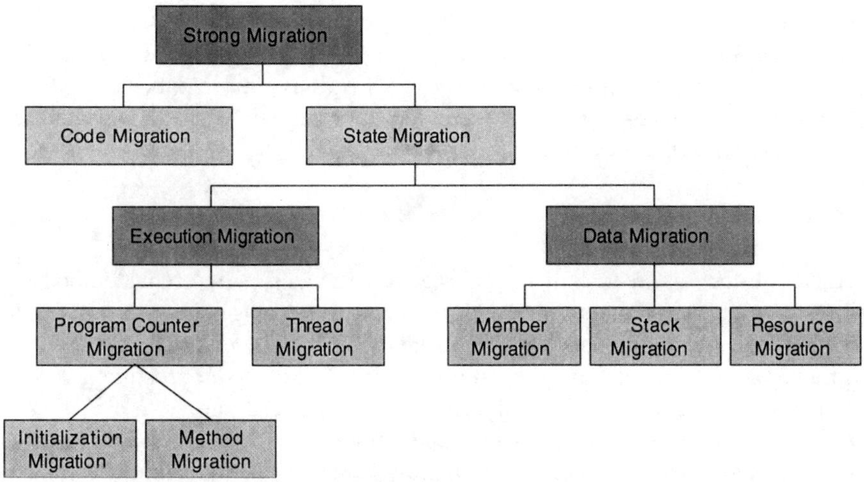

Figure 3: Classification of the different aspects of migration.

1.2 Program counter migration

Program counter migration occurs if the program execution at the destination host continues from the same point where it had been interrupted before. In addition to this kind of migration, we assign a call order to the various methods executed, i.e. the call stack. Since the agent autonomously decides to migrate from one host to the other, unlike migration in load balancing systems, inside the agent code there are some predefined points where migration can take place. These points form a *method call* and are generally referred to with the words *move, jump, fork go* or *migrate.*

1.3 Initialization migration

The achievement of this type of migration is quite common. This technique is a feeble alternative to program counter migration. This means that a mobile agent execution always starts from the same point called *init* which is similar to that used for Java applets. The programmer can carry this method overwrite out in order to branch the agent code towards different locations. For this reason, the programmer must store the agent state and restore it on his own state. Some agent-based systems extend this migration technique with *handlers* called upon before and after a migration occurs, such as in IBM Aglets [2]. At any rate, the process of migration is not transparent.

1.4 Method migration

Like initialization migration, this type of migration is also a feeble alternative to program counter migration. In this case, the programmer agent specifies which method the execution must be resumed in after migration. Different agent entry points are shown as different method definitions. This implies that the code is structured so that it obtains migration towards various entry points. One extension to this kind of approach is that data migration can be passed as a parameter to the next called method, as in the Voyager system [3].

1.5 Thread migration

All the sub-threads created by the same agent must be taken into consideration as far as an agent migration is concerned. The correct restoring of these threads and of their states (running, suspended, blocked, etc.) requires all synchronization variable values (such as semaphores) restore or the state monitor restore, as well as the execution points inside each thread restore. Thread migration makes use of all other aspects, such as program counter, stack, member and resource migration. The main difference between a single-threaded agent and a multi-threaded one is that migration in a multi-threaded system is triggered by a single thread. Therefore, as for all other threads, migration could occur at every point of their code and not only at previously defined points. For this reason problems similar to those present in load-balancing systems occur.

1.6 Member migration

Member variable migration of an agent is very important when state migration of an agent occurs, because this is the place where an agent usually stores its requests and the results obtained. Member migration must be applied to all the objects it refers to (see code migration) and to all concurrent threads (see thread migration.)

1.7 Stack migration

Stack migration involves migration of each method local data to call stack. These data consist of the *variable stack* and the *operand stack* and can be found up to the interruption point. Stack migration depends on program counter migration.

1.8 Resource migration

External resources that an agent can take up until the moment of migration are references to external objects, such as CORBA, EJB, RMI remote objects, open databases, message middleware components, open sockets or local files. Except for local file access, all problems can be reduced to stream unicast or multicast migration.

Let us have a look now at another definition of strong migration, describing it as a migration technique that carries out both code and state migration. State migration requires both execution and data migration. Execution migration is achieved through both program counter and thread migration. Data migration requires stack migration, resource migration and member migration.

Weak migration is an alternative migration technique, if any aspect of strong migration is lacking. We can therefore derive a "stronger than" partial order to say that a migration technique is stronger than the other according to the tree levels shown in fig. 3.

Problems concerning the agent status capture and restore can be found in both strong and weak migration. In distributed applications, process state is captured and sent to some other hosts. The receiving host creates a local process having exactly the same state as the captured process. Status capture can also be used to provide distributed systems with fault tolerance. Programs or processes status is captured at regular intervals and recorded on a permanent device of secondary memory. When the system starts up again, after a deliberate interruption or not, saved information is used to restore the process, that is able, thus, to keep working. Another problem linked to a program status capture is that all information required for the subsequent reconstruction is saved in different memory areas, and it is possible to have access to the program variables only from the inside (i.e. from the language level), while all runtime information lie in lower hierarchical levels. The mechanism of status capture gathers all information from different hosts.

2 Mobile agents migration methods in Java

All the above remarks and those we are going to make hereafter involve the knowledge of Java programming language, which best fits this kind of applications. Java is widely used for mobile agents programming, due to its main features, such as independence from platform, safety, automatic memory management, etc. Furthermore, Java offers a series of flexible mechanisms, such as object serialization, threads, reflection, etc. Though Java does not support status capture, migration process can be carried out through the use of the flexible mechanisms above mentioned, usually available in Java.

There are three main Java-based methods apt to solve the problem of capture and restore of a mobile agent state (fig. 4). In the first approach, called *explicit management*, the programmer must explicitly manage backups in his agents. Backup management consists in storing inside a memory area all the data run by the stack which the agent depends upon. In Java, this memory area is a Java object belonging to the agent state. When the state of an agent is restored, this backup object is explicitly used by the agent in order to make it run again from the point where it had been interrupted before. For example, in applications that implement weak migration in mobile agents, the programmer must usually manage his own program counter. The other two approaches that supply a transparent mechanism are both referred to as *implicit management*. Mechanism is independent from the agent code and is able to capture the agent state (both the data and the runtime state).

These two approaches differ in their implementations: the first approach is implemented through the extension of Java VM in order to make the state

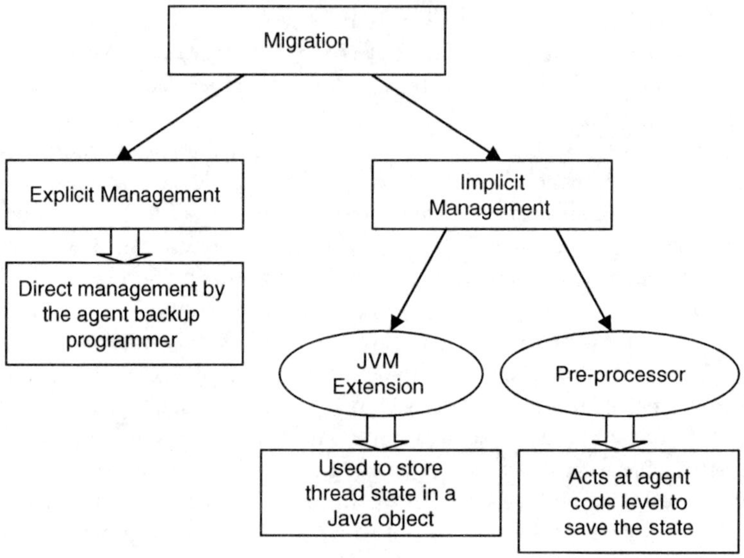

Figure 4: Mobile agent state capture and restore.

of threads accessible to Java agents. This extension enables to easily extract the state of thread and to store it inside a Java object that later can be sent to another machine. This extension, besides, easily enables the construction of a new thread initialized with the previously captured state. At present, some systems are able to capture the state required by Java-based agents through a change of Java VM. The second approach is implemented through the action of a pre-processor that operates on the source code of the agent in order to insert instructions for the restoring of thread state in a backup object. The advantage of this kind of approach is that it does not need to modify Java VM. When an agent requires a snapshot of thread state, it must simply use the backup object created by the pre-processor code inside the agent code. Restore is obtained through the execution of a different version of the agent code (produced by the pre-processor) that restores the stack and the local variables initialized by the values stored in the backup object. This is the basic idea of the mechanism proposed in Ref. [4].

2.1 State capture

Java being an object-oriented programming language, each Java program state includes the state of all the objects existing at the moment of the capture, the method call stack resulting from the program execution, and, finally, the program counter. Java is, besides, an interpreted language that needs an interpreter (the Java VM) to execute its programs. The method call stack and the program counter are indeed located in VM; thus, it would be enough to access the information contained in it to acquire them. Using then Java object serialization, we can obtain the state of all the objects existing at the moment of the capture. That is what is necessary to capture the status of a Java program at the language level.

The serialization of a Java object is a simple method that does not take into account the state of all the objects existing inside an agent. This state includes the values of each object variable (for example class and instance variables), representing its inner state, and information concerning the kind of object. Using object serialization, great part of information (all information at language level) required to restore the agent state can be captured. What is lacking, however, is information located in VM, i.e. the method call stack containing all values of each method local variables, and the program counter current value. Thus, the use of a pre-processor is provided. By using it, we can use the user's Java code, adding to it additional code carrying out the present state capture and restoring the state, so that the agent itself is able to keep running in the destination host. This is obtained through an analysis of the original program code by using a syntactic analyser based on Java, generated by the Java 1.1 JavaCC tool. The pre-processor uses and modifies the syntactic analyser from which the new code springs.

Since the additional code introduces time and space penalties, we will work just on those code parts that need it and also additional code will be executed only if necessary (that is when the state capture occurs). We will now describe a special method that is responsible for all the methods local variables and the program counter value saving: when they are forwarded, the normal execution flow stops immediately. An error can be detected by a *catch* clause of a *try* instruction. If the error is not removed, it spreads all over the method call stack. This is automatically done by the *exception/error handling* mechanism of the Java VM, whose purpose is to handle errors and exceptions. This behaviour is exploited to go through the method call stack and save all local variables of each method present at that moment on the stack. The approach consists of the following steps:

a. the method beginning the process of status saving locates an error;
b. pre-processor puts an encapsulated *try-catch* directive for each method which should begin the state saving in order to save the above-mentioned local variables;
c. after the execution of the code concerning local variables saving in the *try-catch* statement, the error is forwarded again. Thus, each method in the stack in turn removes the error that brings to the execution of the saving code of the variables present in such method.

The pre-processor, exploiting the Java error handling mechanism, goes through the method call stack putting in each method the part of code for that method variables saving. Thus, *Myprogram* class code of fig. 5 is transformed in the code shown in fig. 6.

Using an error rather than an exception to achieve the saving of the state has the benefit that errors must not be made clear somehow in the method. In order to save

```
Class Myprogram { // variables definition
 ...
 public void mymethod (int I, real j, MyObject   m)
 { int  k; /* k can be any value */
           /* in any part of the program */
 Hashtable h;
 ...
 saveState(); ... /* saves the value of I, J, K, m, h */
 if  (k=5)
 { Vector x = new vector(); ... saveState(); ... }
 /* saves the x value */
 ...
 int v = 10; /* v is a local variable in this method */
 ...
 saveState(); /* saves the v value of v */
 ...}
 }
```

Figure 5: Class using state saving.

```
Class Myprogram {
// variables are saved by serialization
...
 public void mymethod (int I, real j, MyObject m)
 { int k;
 Hashtable h;
 ...
 try { saveState( ) ; }
 catch  (Migration  mig)  {  save(h);  save(k);  save(m);
  save(j); save(I);
 throw mig; } ...
 if ( k==5 ) { Vector x = new Vector ( ); ...
 try { saveState( );}
 catch  ( Migration mig ) {save (x); save(h); save(k);
  save(m); save(j);save(I); throw mig; } ...
 } ...
 int v=10;
 ...
 try { saveState( ); }
 catch  (Migration  mig  )  { , save(v);  save(h);  save(k);
  save(m); save(j); save (I);
 throw mig; }
...}
```

Figure 6: Transformed *Myprogram* class.

the values of all local variables we use a special *save* object that the pre-processor has put in the higher class of the Java-based agent. In addition to this, all the methods that should be included in the state saving are endowed with this special object. Thus, the method can be called a local method. Because of this, all the relevant methods signatures must be used. Unfortunately, this leads to some problems in the inheritance tree when such methods are part of an interface that must be implemented by the class. Our solution to this problem is to generate a new interface that incorporates the method signature tool. After having saved and restored the stack, all information about the state is kept in a special *save* object. Since this object is inside the highest-level class, its value can be saved by a normal serialization mechanism. Depending on the purpose of the state saving mechanism, the serialization information can be written in a file or in a socket of the network.

2.2 State restoration

Capturing the state of a Java-based running agent is just being halfway towards reaching our goal. We must also be able to restore a *program state* from the state information formerly saved.

From the program point of view, the control flow should be carried on immediately after the directive starting the state saving process. This task requires the reconstruction of the program graph and of the object states, the reconstruction of the method call stack, and the restore of the local variables values concerning each method on the restored stack. Most of the program state can be automatically reconstructed from the serialization information run by the de-serialization process provided by Java. This process provides an object graph which shows the same connections and the same features of the object state as those provided by the object graph representing the program at the time of serialization. What is lacking is the method call stack that has not been automatically reconstructed.

2.3 Method call stack reconstruction

Since the *save* object (which keeps the relevant information) is part of the program object graph, we can use that information to fill all the method local variables with the correct values once the method call stack has been recreated. In order to do that, we need but retrieve all the relevant methods following the same order they had on the stack when the state capture occurred. To prevent the code re-execution of already executed methods, we must skip over all those parts of each method code that had already been executed before the state capture occurred. Thus, we introduce an *artificial program counter.* It states, for each modified method, the already executed directives so that they may be skipped when the method call stack is being reconstructed. It is not necessary to modify the artificial program counter after every instruction, as the next directives that have not started the state saving can be considered as a single composed directive. All instructions previous to and next to the instruction set by the artificial program counter are treated as instruction blocks. Each block of mixed instructions is controlled by a conditional *if* that checks if the artificial program counter states that this group of mixed instructions should be executed or skipped over.

2.4 Local variable values set up

The method call stack has been restored as shown above. Now each method local variables must be changed according to current values (that is with the values

The code:

```
real I ; int j = 7 ; Integer x = new Integer(5);
```

is transformed in:

```
// init I
real I = 0.0; if ( restart ) I = so.restore (i);
// init J
int j = 0; if (restart) j = so.restore(j); else j = 7;
//init x
Integer x = null;
if ( restart ) x = (Integer) so.restore(x);
else x = new Integer (5)
```

Figure 7: Local variables transformation.

saved in the *save* object). To obtain that we insert a code that modifies each variable giving it its correct value. There are two ways for a variable to assume its own value: the original value (the starting one) provided by the programmer in case of a mobile agent that normally sets up, and the value stored in the *save* object in the case of an agent launched again, i.e. that continues to run from its former point of interruption, showing inside each local variable the value properly stored. To satisfy the Java compiler, all variables are initialized to a default value. The assignment of updated value is made in the context of an *if* directive. Fig. 7 shows this transformation.

2.5 Thread recovery

Unlike what already done in the case of object serialization, it is not possible to transfer execution thread state in Java. Since each Java program is executed like a Java VM thread, the converter is able to save each single thread state. To save the state of all the threads in the program, we simply use a new *save* object for each thread that keeps the information contained in the method stack of the associated thread. To the restart, all the threads existing at the time of the state saving are newly created and their runtime information is read by their *save* object. To save each thread runtime information is simple, but other problems require our attention: since threads run currently, we cannot foresee when a thread that requires the state saving will start it, nor which other threads state will be in the same moment. As to all other threads, the request for state saving might occur at every instruction, but that turns out to be both impossible and inefficient. Thus, the programmer is provided a pair of new methods called *setflag*( ) and *allowgo*( ) (figs. 8 and 9).

Each thread should execute *allowgo*( ) before asking for state saving. *Allowgo*( ) method controls that another thread has not required state saving. Otherwise, it comes back immediately not to block the current thread, until all execution threads have called on the *setflag*( ) synchronization method. Thus, the state saving occurs only if all the running threads have called on the *setflag*( ) method.

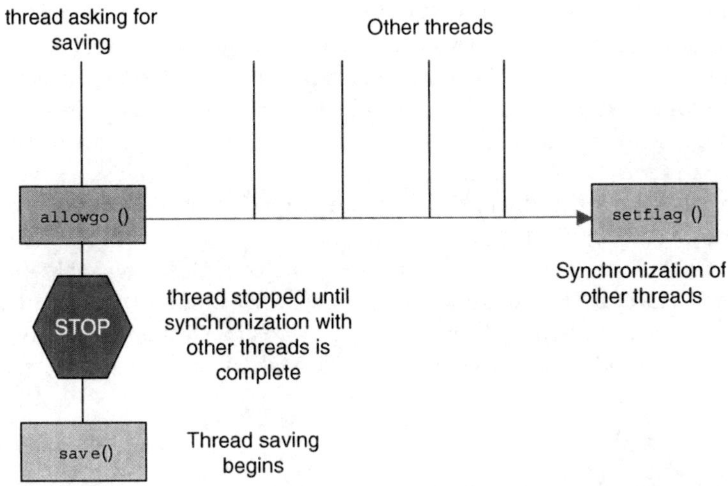

Figure 8: Methods for thread synchronization.

```
Setflag( ): public syncronized void setflag ( )
{ flag[ I ] =1 ; resume( ) ; }
// flag is an array of integer, inizialized to 0

allowgo( ): public syncronized void allowgo( )
{ for ( int I =1; I<=N; I++)
{ if  ( flag[ I ] = 0 ) suspended ( );  }  }
// N is the number of the currently running thread
```

Figure 9: Code for thread synchronization.

3 Mobile agent itinerary planning

A very important aspect concerning any mobile agent-based application is the one concerning the planning of the itinerary to be covered in order to achieve the goal. Planning the itinerary means to decide which Internet nodes the mobile agent must go to. It can be done by the following different criteria, for example by setting an *a priori* list of sites that the agent will visit sequentially during the execution, or by instructing it with selection criteria. That will enable it to follow one path rather than another in order to complete the task, according to the operations performed in a certain place.

Actually the problem is more complex than it seems. As a matter of fact, increasing the applications complexity, and therefore, equally, the competencies required of an agent, itinerary planning proves to be problematic as to network traffic and execution times, especially in the case of applications using a great number of agents.

The use of few agents has the advantage of causing less traffic in the network and conserving bandwidth. This occurs whenever an agent travels on a fixed set of nodes to be visited. On the other hand, however, the number of agents used for the execution of a task seriously affects execution times. Using a limited number of agents, while having positive implications on the one hand, could bring to an increase in the total time of execution on the other. At the same time we must be aware of the fact that the more agents we use, the more the total cost of router forwarding rises.

The problems just mentioned have been the object of remarkable interest to researchers who have tried to find a balance between the various parameters at stake, minimizing or maximizing them and trying to look for an almost ideal situation for the execution and achievement of the specific goals of various applications.

3.1 MAP vs. TSP

A good approach for the solution of these problems can be found in Ref. [5], where they are dealt with in the perspective of mobile agents technologies in information retrieval systems.

MAP (Mobile Agent Planning) approach is different from the more commonly known in literature such as TSP (Travel Salesman Problem), though the trend has been to deal with the two issues in an almost equivalent way. The two approaches are actually different; in fact TSP deals only with the optimization of forwarding time for a given number of agents, while MAP intends to minimize the whole time of execution to complete information retrieval. Another difference between TSP and MAP is that in the latter the agents' possibility of visiting a node more than once is scheduled, whereas in TSP it is not scheduled.

In the lower part of table 1 we have introduced some functions that we will use later in the algorithms dealt with. Hereafter we will explain some of them in detail:

- The agent's tour (Tour(A_i)): Tour(A_i) is the tour of A_i agent; it includes all the nodes to be visited by the agent to achieve its task;
- Shortest latency ($L_s(h_i, h_j)$): it returns the shortest latency time between the nodes h_i and h_j, resulting from an evaluation of the shortest routes among all nodes pairs;
- Tour time (TourT(S)): it returns the execution time of a tour. We use two formulae for the calculus which refer to the cases of a single tour or more chained tours and are reported below:

$$\text{TourT}(S) = 2 \times L_s(H, h_1) + \text{Comp}(h_1)$$

$$\text{TourT}(S) = \text{TourT}(S_1) + \text{TourT}(S_2) - L_s(H, \text{First}(S_1)) - L_s(H, \text{Last}(S_2)) + L_s(\text{First}(S_1), \text{Last}(S_2))$$

The two cases that the above-mentioned formulae refer to are graphically reproduced in fig. 10(a) and (b) below.

Table 1: Symbols and keys.

Symbols	Description
N	Number of nodes, except starting node
R	Number of mobile agents used for a task
H	Home node
δ	Time of execution for the achievement of a task
$h_1, h_2, \ldots, h_n$	Node identifier
$A_1, A_2, \ldots, A_r$	Agent identifier
Tour	Series of nodes visited by the agent
Tour(A_i)	Agent A_i's tour
Comp(h_i)	Computation time at h_i node
$L_s(h_i, h_j)$	Shortest latency time between node h_i and h_j
Union($S_i, \ldots, S_j$)	Chain of tours, where $S_i, \ldots, S_j$ represent tours
TourT(S_i)	Forwarding time, namely the execution time, for S_i tour
First(S_i)	First entry of S_i tour, i.e. i_j
Last(S_i)	Last entry of S_i tour, i.e. i_k

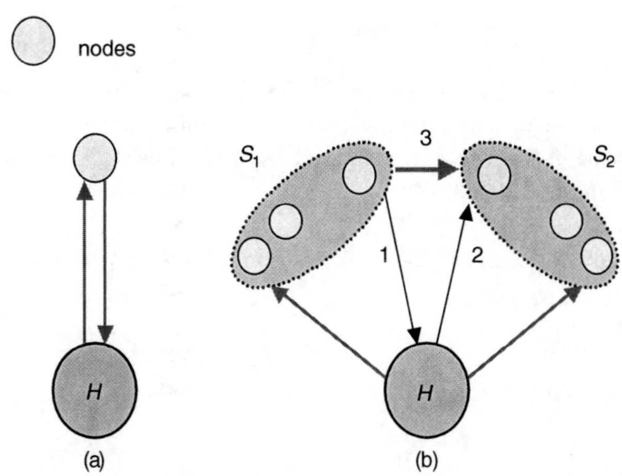

Figure 10: Tour time computation.

In the first case the tour consists in the visit to a single node h_1 and the tour time is calculated as the sum of the latency times of come and go back from the home node to the other node plus the computation time on the node.

In the second case we assume that the tour results from joining the S_1 and S_2 tours. In that case the tour time is calculated as the sum of S_1 and S_2 tour times minus the latency times between S_1 home node and entry node and S_2 home node and entry node, in their turn added to the latency time between S_1 entry node and S_2 exit node.

3.2 MAP problem definition

In the definition of our problem, first we assume to know the history and the statistics of the network which we are working in, through a convenient monitoring of the same network. Therefore, we can carry on our proper definition of the problem:

Suppose we have $n + 1$ nodes ($H, h_1, h_2, \ldots, h_n$), where H is the starting node or home node. Each node is combined a computation time coinciding with that necessary to an agent to carry out its task on that node. Suppose we also know the latency times concerning the agents movements for each couple of h_i and h_j nodes. The computation time at H node, indicated as t_h, is supposed to be null. The problem is to minimize δ, the execution time and the number of mobile agents R, in order to successfully complete the task.

As better explained further on, the execution time (δ) can be defined as the longest forwarding time (TourT) among all the nodes that can be visited by an agent, which means that

$$\delta = \text{Max}\{\text{TourT}(h_i)\} \quad \text{with } 0 < i < n \tag{1}$$

Our purpose is to choose the series of nodes that every agent must visit, the minimum number of necessary agents, and the minimum execution time for a task achievement. Such problem, called CE-MAP (Cost Effective Mobile Agent Problem), is defined as follows:

$$\text{Minimize } wr + \Sigma_{1<i<r} \text{TourT}(\text{Tour}(A_i)) \leq \delta \quad 1 < i < r \tag{2}$$

Subject to:

$$\text{TourT}(\text{Tour}(A_i)) \leq \delta \quad 1 < i < r \tag{3}$$

$$|\cup_{i=1,\ldots,r} \text{Tour}(A_i)| = n, \quad \text{Tour}(A_i) \cap \text{Tour}(A_j) = \varnothing, \quad i \neq j \tag{4}$$

$$\delta = \max(\text{TourT}(h_i)) \quad 1 < i < n \tag{5}$$

In the expression (2), w is given a higher value than the total forwarding cost, namely the sum of the tour times of all the agents appearing in the second term of eqn (2). Therefore, the expression (2) has two goals: the minimum number of agents on one side and the minimum cost of forwarding on the other side.

Formulae (3), (4) and (5) are subject to this goal function. In the expression (3), each agent's forwarding cost is smaller and smaller than the minimum one, represented on the whole by the δ minimum execution time. This means that a given task can always be accomplished inside this minimum value, which is determined by the node presenting the highest value resulting from the sum of latency time and computation time. The expression (4) requires instead that each node undergo the process just once.

3.3 MAP problem solution

Let us assume we have at our disposal a network monitoring system able to give us necessary information such as the latency times and the bandwidth of host links, so as to be able to know all the latency times between existing connections. Let us suppose, furthermore, that latency times inside a network are variable, and that the agents are created in the H *home* node only and that they are no way cloned in other nodes. We do not take into consideration, besides, any probability of succeeding in the various nodes data restore.

The attempt to determine the overall δ minimum time goes through the knowledge of latency times of all the possible pair of nodes. To establish these, however, is not a simple task, because it is necessary to determine all possible combinations of connections between all the possible pair of nodes. Without this kind of information, we have no chance to determine the minimum time of execution. In order to obtain latency times of all the pair of nodes, the algorithms, which we will discuss later, before going into the main body, pass through a pre-algorithm to determine the shortest route among the pair of nodes and to build a network graph based on the shortest latency times.

Fig. 11 shows a network feasible configuration that can be divided into two parts.

The value on each corner represents the latency time expected for the connection between the couple of nodes connected by the same corner. The value on each node corresponds to the computation time on that node. The minimum execution time is never shorter than 50 ms, which corresponds only to the value of the h_1 node, and which is given by the sum of latency time of come and go back from H and computation time ($10 + 30 + 10 = 50$ ms). The other two nodes, h_2 and h_3, have instead forwarding costs equivalent to 21 ms and 25 ms, respectively.

In our example the minimum number of agents to be used is two. The first agent covers the h_1 node as shown in fig. 11(b), while the second agent covers the h_2 and h_3 nodes as shown in fig. 11(c). The total time of execution is 88 ms ($= 50 + 38$).

Therefore, in conclusion:

$$\mathrm{Tour}(A_1) = (h_1) \text{ and } \mathrm{Tour}(A_2) = (h_3, h_2)$$

In this example the presence of the pre-algorithm we were talking about, used to determine the shortest routes between nodes couples, finds its relevant

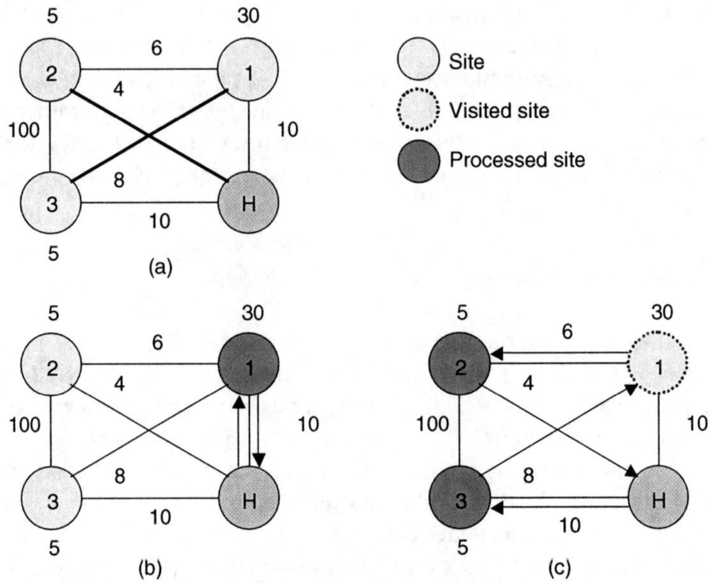

Figure 11: Feasible network configuration for the calculation of the shortest latency time between two nodes.

expression in the second agent that, in the passage required by h_3 and h_2 nodes, prefers to pass through h_1 node, rather than use direct connection which, as we can see from the value on the corner connecting the two nodes, is very costly as far as time is concerned. Therefore, the distance (route) covered by the agent is (H, h_3, h_1, h_2, H) rather than (H, h_3, h_2, H). After reckoning up, we can see that the total time of the first route amounts to 38 ms, vs. the 128 ms of the second one.

After having provided the acquisition of the minimum time of latency between connections (that we indicate as $L_s(h_i, h_j)$), the algorithm proceeds as follows:

- nodes are distributed in a decreasing order in a list according to their forwarding time TourT(h_i);
- δ is initialized with a value equal to the execution time of the first node in the list;
- the network is divided into many parts and nodes are grouped in a way that the execution time associated to each of these parts does not exceed the initialized δ value, and a forwarding route is created for each partition;
- a TSP algorithm is executed to optimize every forwarding route;
- an agent is allocated and sent to every partition.

The developed algorithms are called BYKY1 and BYKY2 (figs. 12 and 13). They pursue the same goal, but use a slightly different partition method. Their main difference consists in the fact that the second algorithm is more dynamic than the first. It is convenient to remember that at the end of both the algorithms,

Nodes sorting:

1. sort nodes from farthest to nearest
2. Put the result in a list ($h_{a1}, h_{a2}, \ldots, h_{an}$)
3. Save the longest route in δ (i.e. δ = TourT(h_{a1})

Agents planning:

4. Repeat for all A_x Agents with $x = 1,2,\ldots,n$ // start loop 1
   - 4.1. Repeat for all h_{ak} nodes with $k = 1,2,\ldots,n$ // start loop 2
     - 4.1.1. If h_{ak} node has not been marked and if, adding to the route of A_j agent h_{ak} node, this does not exceed the δ value,
     - 4.1.2. Then add h_{ak} node to the A_j route
     - 4.1.3. and mark the h_{ak} node // End loop 2
   - 4.2. If every agent has been assigned a route
   - 4.3. then exit loop1 // End loop 1

Agent's optimisation:

5. Call optimisation function for each agent (OPT2)

Figure 12: BYKY1 algorithm.

a TSP algorithm called OPT2 is used. Its task is to produce a forwarding route that asymptotically tends to the optimum situation.

3.4 VMAS (Visual Mobile Agent System with itinerary scheduling)

In the previous section we have dealt with the problem of mobile agents migration with a particular respect for the solution of the problems caused by such migration (number of agents to be used in every application, network traffic, total execution time of the goal, etc.)

We will now refer to a solution which, even if a less rigorous one, has the merit of being nearer to the level of the user who adopts a mobile agent-based technology. We will in fact propose a real application that, through a user-friendly interface, allows the single user to intervene during the itinerary scheduling of mobile agents.

The proposed application is called VMAS [6] and is a Java-based system consisting of three main components called *Agent Manager, Meta-Service Server* and *MA Builder. Agent Manager* controls the sending, reception and execution of agents, as well as any possible communication with other agents. *Meta-Service Server* offers an itinerary scheduling system for agents. Its function is very similar to that of Yellow Pages. *MA Builder* provides the agent's actions and can manually intervene in scheduling its itinerary.

With respect to this last point, MA Builder provides a visual way that enables the user to manually intervene on the agent itinerary scheduling. Using the functionalities called "New Node" and "Node Connect", the user can easily include nodes in the agent's migration route. On the other hand, the user can also define

Nodes sorting:
(*see BYKY1 algorithm*)

Agents planning:

4. Repeat for all A_j agents with j= 1,2,...,n // Start loop 1
   4.1. Initialization of agent A_j travel
   4.2. Select among h_{ak} nodes the nearest unmarked
   4.3. Mark the selected h_{ak} node
   4.4. Repeat endlessly // Start loop 2
      4.4.1. Create a list of unprocessed nodes in an increasing order as regards the distance between nodes with the chosen h_{ak}
      4.4.2. Put the result a standard list (h_{b1},h_{b2},...,h_{bm})
      4.4.3. Keep trace in previous_k of the h_{ak} node index chosen at point 4.3
      4.4.4. Repeat for all h_{bk} nodes of the list created at point 4.4.2. with x= 1,2,...,m // Start loop 3
         4.4.4.1. If adding to the A_j agent route the h_{bx} node this remains less than δ, then:
            - Add tothe A_j route the h_{bx} node and mark the h_{bx} node
            - Find k value so that $a_k = b_x$ (set next starting node)
            - Exit loop 3 // End if
      4.4.5. If the next node to be visited is the node just processed (there are no more nodes to be processed), exit loop 2
   4.5. If there are no more nodes left for the agents exit loop 1

Agent's optimisation:

5. Call optimisation function for each agent (OPT2)

Figure 13: BYKY2 algorithm.

the conditions for the achievement of the migration between a generic couple of nodes (for example he can decide that agent migration from node A to node B is achieved only if the value of a Boolean variable x is TRUE).

In manual itinerary scheduling, though, the user usually intervenes disregarding the costs concerning the agent's travelling in the net. Nevertheless, thanks also to the knowledge of the environmental situation of the net, it is possible to carry out an itinerary scheduling fit for an optimum situation in terms of total costs. It is obviously quite difficult if not impossible to achieve such an optimum situation just dealing manually with this kind of scheduling.

That is why VMAS includes, besides the manual solution, an automatic solution as well.

3.5 Automatic itinerary scheduling

The method proposed for the automatic itinerary scheduling is called *hybrid scheduling*. First of all, it is based on a twofold classification of the nodes through which the agent must pass: *essential nodes* and *non-essential nodes.*

Essential nodes are those through which a mobile agent is obliged to pass through during its travel. Non-essential nodes are the ones that are not given priority during the agent's travel.

Another classification made in explaining the field of hybrid scheduling, is the one concerning all possible cost factors. It is based on elements such as freedom from time constraints and travel order between two nodes:

- TI-OI (Time-Independent, Order-Independent) $f_{\varnothing}$ (e.g. Service node log cost)
- TI-OD (Time-Independent, Order-Dependent) f_o (e.g. Connection between two nodes cost)
- TD-OI (Time-Dependent, Order-Independent) f_t (e.g. Number of MA resident in a service node)
- TD-OD (Time-Dependent, Order-Dependent) f_{to} (e.g. Transmission speed between nodes)

According to this classification of cost factors, the formula of the cost of the distance covered between two nodes, called node i and node j, can be represented as follows:

$$\text{Cost}^{ij} = \text{Cost}^{ij}_{\varnothing} + \text{Cost}^{ij}_{t} + \text{Cost}^{ij}_{o} + \text{Cost}^{ij}_{to} \qquad (6)$$

Itinerary hybrid scheduling is developed in two steps: the first production of the route and the incremental change.

First production of the route: This part is dealt with before the mobile agent forwarding. In this part a low-cost initial route is created according to the cost functions assigned by the user. The algorithm for the determination of the optimum initial route is similar to the TSP one. Since the mobile agent must reach every essential node, the optimization of the route between essential nodes must not take into consideration order-dependent cost factors. Hence, formula (6) can be reduced as follows:

$$\text{Cost}^{ij} = \text{Cost}^{ij}_{o} + \text{Cost}^{ij}_{to} \qquad (7)$$

We should remember, though, that as far as the generation of initial route of essential nodes is concerned, the calculation must include all cost factors, as we have done in eqn (6).

Incremental change: This part is dealt with after the mobile agent forwarding. The aim is to deal with the way in which the agent itinerary must be adjusted during its travel according to the state of the network at the moment of migration. Fig. 14 shows an example that describes how changes are made.

Let us suppose the agent's initial route is 1→2→3→4→5→6→7→8. When the agent is in node 3 and must leave it, if it notices that the state of node 4 is different from the one it had at the moment of the route initial creation, the route is changed by selecting a new node to visit. Let us suppose that node 7 is selected as the next one, we need to verify if the change is optimal. The decision is taken making a

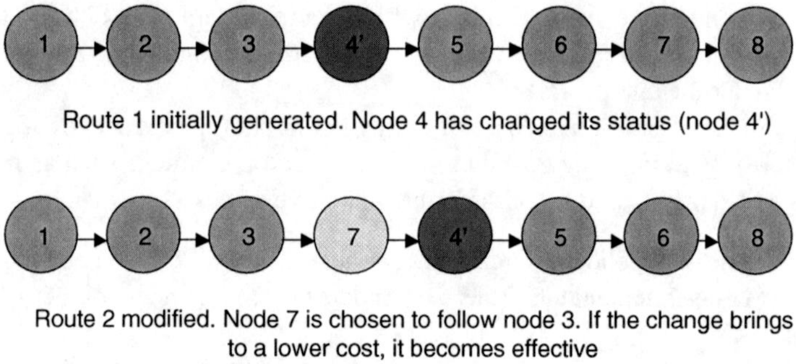

Figure 14: Incremental change.

comparison between the old route cost (3,4,5,6,7,8) and the new one (3,7,4,5,6,8). If the new solution cost is lower than the original solution cost, we can go on with the acquisition of the new route in the agent's itinerary, otherwise the original route is not changed. It is worth noting that in the route incremental change the cost function is only used to detect the differences between the old and the new route; therefore, the TI-OI ($f_{\varnothing}$) factor is not taken into consideration. The cost function thus assumes the following form:

$$\text{Cost}^{ij} = \text{Cost}_t^{ij} + \text{Cost}_o^{ij} + \text{Cost}_{to}^{ij} \quad (8)$$

References

[1] Illmann, T., Weber, M., Kargl, F. & Kruger T., Migration of mobile agents in Java: Problems, classification and solutions, *Proc. of the International Symposium on Multi-Agents and Mobile Agents in Virtual Organization and E-Commerce,* ICSC, Academic Press: Canada, pp. 362–369, 2000.

[2] Lange, D.B. & Chang, D.T., Programming Mobile Agents in Java, IBM Aglets Workbench, IBM Corp. White Paper, September 1996.

[3] Voyager homepage, http://www.objectspace.com/voyager/prodVoyager.asp

[4] Wang, H., Zeng, G. & S. Lin, A Strong Migration Method of Mobile Agents Based on Java, The Sixth International Conference on Computer Supported Cooperative Work in Design, IEEE, 2001.

[5] Baek, J.W., Yeo, J.-H., Kim, G.-T. & Yeom, H.-Y., Cost Effective Mobile Agent for Distributed Information Retrieval, 21st International Conference on Distributed Computing Systems, IEEE, 2001.

[6] Chang, J.S. & Chang, C.Y., A visual mobile agent system with itinerary scheduling, *Proc. of the 4th Int. ACM Conf. on Autonomous Agents*, Barcelona, Spain, pp. 167–168, 2000.

Communication

Alfieri Cimadori, Giacinto Giambalvo, Miriam Miceli,
Teodoro Ricciarello, Gabriele Santangelo, Salvatore Sollami
and Alessandro Genco
DINFO – Dipartimento di Ingegneria Informatica Università degli Studi di Palermo

1 Introduction

The ability to communicate is an essential characteristic of mobile agents. Significant efforts have been made to allow communication among mobile agents, defining a common semantic background and *ad hoc* languages for the exchange of information, such as ACL (agents communication languages) or KQML.

The creation of facilities for mobile agent communication has nevertheless a problem at a lower level of abstraction, i.e. the reliable delivery of a message when the mobile agent's mobility schemes are *a priori* unknown.

If, in fact, mobile agents can freely move from one host to the other according to some *a priori* unknown migration scheme, the possibility of reliably delivering information is linked to the one of determining the position of the mobile agent, and to ensure that the message reaches the mobile agent before it moves again.

Current mobile agent systems are based both on conventional communication devices like socket and remote procedure (or method) call, and on implementation of their own devices for the exchange of messages.

A typical mobile agent paradigm does not make use of a direct communication link, but schedules local access to the resources of a remote server. It could be objected that communication with a remote agent is not important and that a mobile agent platform should focus on locally exploited mechanisms of communications, in order, for example to access to the server or to communicate with the agents staying in the same site. Many mobile agent systems provide mechanisms for local communication, using some sort of abstract meeting (as initially proposed by Telescript), or event announcements for group communication, or, more recently, *tuple spaces.*

Different common settings exist though and it is necessary to guarantee communication also among agents in different sites. Let us imagine a master agent generating a number of slave agents and afterwards introducing them in the network to achieve some kind of cooperative operation, for example to restore some information. At a certain point, the master could want to stop the slave agents execution, for example because the piece of information required has been already found by one of them and it is therefore advisable to prevent unnecessary waste of resources. Otherwise, some parameters ruling agents' behaviour could be changed, because the environment that had determined their creation has changed. Or rather, slave agents could want to check if the master agent is still alive by executing some detection methods.

The supposed scenario needs the presence of a mechanism for the exchange of messages, ensuring that the message is effectively delivered (at least once) to the recipient, independently of the source movement and of the communication recipient. Typical delivery schemes have the problem that a message to an agent that is migrating at the moment of delivery can easily get lost. In order to clarify the concept, we will discuss hereafter two different message delivery approaches: *broadcast* and *forwarding.*

A simple broadcast scheme assumes a scattered tree of the network nodes through which a message can be sent from any node. This node broadcasts the message to the next ones that will send it in turn to their next ones and so on until it arrives to leaves. This method does not ensure the message delivery when the agent is travelling in the opposite direction. If the agent is being transferred while the message is spreading in the opposite direction, they will cross in the channel, and delivery will never take place.

In a simple forwarding scheme the agent is meant to provide a pointer to a well-known location, called *agent home-place.* During its migration, the agent must inform the *agent home-place* of its new location, in order to enable new communications. Anyway, all the messages sent during the agent's migration and before updating its position are lost. Forwarding has another disadvantage: every time the agent migrates, communication with the *agent home-place* must be asked for. In some cases this can undo mobile agents advantages and restore centralization. In the presence of many mobile agents in the same host, this scheme can lead to a considerable network traffic all around the *agent home-place*, and to slow down performances if latency time between the mobile agent and the *agent home-place* is high. Finally, this approach is intrinsically difficult to apply when disconnection operations are needed due to that "umbilical cord" with the *agent home-place.*

Current mobile agent systems make use of different communication strategies. OMG MASIF standard details only the interface enabling the call and the location of an agent in different platforms. Some systems, notably Aglets and Voyager, make use of forwarding by associating a proxy object playing the role of *agent home-place* to each mobile component. Others, like Emerald, make use of forwarding and then resort to broadcast when the recipient cannot be found. Finally, some systems, e.g. Agent Tcl, provide mechanisms based on common remote procedure (or method) call, and leave to the application developer the management of a delivery failure.

A related subject is the supply of a mechanism for reliable communication to a group of mobile agents. Group communication is a useful program abstraction to deal with functionally connected mobile agent groups to whom the same part of information must be sent. Many mobile agent systems, notably Telescript, Aglets, Voyager and Jade, allow the possibility of multiple message sending only within the context of a single runtime support. Finally, Mole provides a mechanism for group communication that, however, assumes that the agents are stationary during the exchange of information.

2 Effective communication

As already said, simple mechanisms for message delivery, such as broadcast in a scattered tree and forwarding, can fail when agents are in transit or moving quickly. Generally speaking, in order to solve these inconveniences, agents must be confined to areas where they cannot go out from without receiving a copy of the message. For example in the above-mentioned broadcast mechanism, the case of the agent moving in the opposite direction than that of the message on a bi-directional channel has been considered. In this case, if the message is still present in the agent's destination node, it can be delivered when the agent reaches the node. So, the message is kept in the node until it has been delivered. Although this simple extension would guarantee delivery, it is not reasonable to expect nodes to hold messages for arbitrary periods of time.

A solution imposing a tight bond on a message storage time in a node would lead to entrap the agent in a region of the graph so that, wherever it transfers, it cannot escape receiving the message.

The first algorithm shown assumes that the network of nodes and channels is known; it presumes as well that a single message at a time is present is the system. In such a context, just one delivery of message is ensured without changing the agents' behaviour both as far as mobility is concerned and in accepting messages. We will show later an adjustment for the delivery of multiple messages.

Though these algorithms enable a reliable delivery of messages, the assumption that the whole network graph must be known in advance is unreasonable in situations where mobile agents are used. There are improved algorithms allowing the graph to grow dynamically as the agent moves. In order to be clearer, we will deal with this adjustment in two phases: we will first make messages originate in a single node, and later we will make it possible for any node to send a message.

2.1 The logical model

The logical model is the typical network graph where nodes are the agents' locations, and vectors are the FIFO channels through which the agents can migrate and messages be broadcast. FIFO assumption is crucial for the algorithm's correct execution.

Let us assume a connected network graph (i.e. let us assume that a route between each couple of nodes exists), but not necessarily totally connected (i.e. a channel between each couple of nodes does not compulsorily exist). In a typical IP network, all nodes are logically connected: it is the case of a network of sub-networks connected through gateways which an agent must necessarily use to enter or leave a sub-network, as shown in fig. 1.

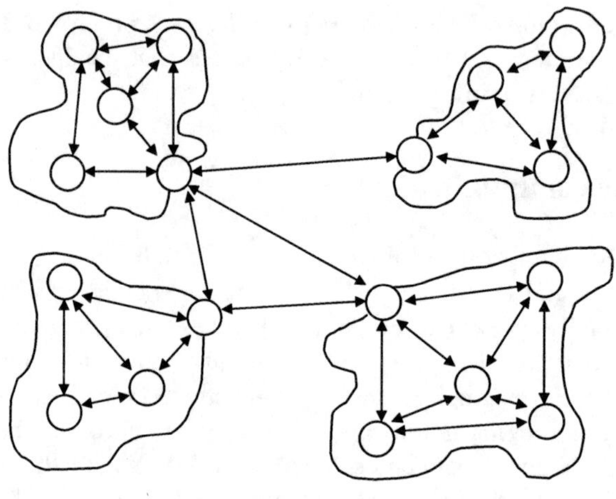

Figure 1: A network of sub-networks. Agents can enter and leave sub-networks only through gateway servers.

Let us consider also that the mobile agent's server keeps traces of the agents it is hosting, and that it provides some basic mechanisms to deliver a message to an agent, for example by calling one of the methods of the agent's object. Finally, let us assume that each agent has a single global identifier to be used to direct a message to the agent. These last assumptions are acceptable since they are already fulfilled by most mobile agents platforms.

2.2 Delivery of a single message in a static network graph

The algorithm works by associating the *flushed* state or the *open* state to every channel entering a node. At the beginning all channels are *open*. When a message reaches a channel, the state is transformed in a *flushed* one, and all the agents in that channel are obliged to leave it. When the message arrives for the first time in a node, it is kept and spread on all the outgoing channels, starting the propagation process. The message is also delivered to all the agents in the node. All the agents that reach a node through an *open* state channel receive a copy of the node's message. When all the channels converging into a node are in the *flushed* state, the node is not asked anymore to deliver the message to every arriving agent, so the copy of the message is deleted and all channels are automatically set to *open*.

2.3 Delivery of a multiple message in a static network graph

A simple adjustment of the previous algorithm for the delivery of a multiple message should require that a node waited till the end of the message delivery and coordinated itself with the other nodes before sending another one, so as to ensure the presence of a single message in the system.

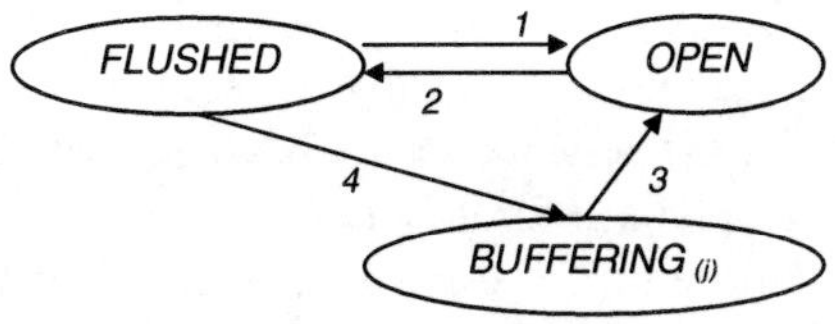

Figure 2: State transitions diagram for multiple message delivery on static network graph message delivery.

In an approach based on multiple messages contemporarily present in the system, the node where the message originates assigns it an exclusive sequence of numbers. In practice, the sequence of numbers allows the nodes to manage multiple applications of the algorithm, in order to include both the case of a single source broadcasting a sequence of messages and that of multiple sources broadcasting at the same time.

There is a problem concerning a new message arriving before the current message has been processed. In this case, the channel is already in *flushed* state, but not all the other ones are. In order to deal with such a case, *buffering* state is introduced, as shown in fig. 2, where every message arrived on a *flushed* channel is put in a buffer to be processed later (see transition 4 in the figure). *Buffering* channel is not taken into consideration when transition from *flushed* to *open* state is determined. When transition has finally occurred (transition 1 in the figure), all *buffering* channels are put on *open* state (transition 3 in the figure), and messages on buffer's queue are treated as messages just arrived on the channel, and thus processed again. It is possible that, after having processed the first message, the following one causes another *buffering* transition, but the fact that the head of channel is processed ensures progress through a sequence of messages to be delivered.

1	Start:	no channels in *open* state
	Action:	current message = **none**
2	Start:	message $j \wedge$ arrives (current message = **none** $\vee$ current message = j)
	Action:	if current message = **none** deliver, deposit, spread
3	Start:	message j finished processing
	Action:	?
4	Start:	message $i \wedge$ arrives (current message $= j \wedge i > j$)
	Action:	put message i into the buffer

2.4 Delivery in a dynamic network graph

Though solutions hitherto explained ensure delivery in the presence of mobility, the need to know beforehand network adjacent area is unreasonable in the dynamic field of mobile agents. Besides, the delivery system does not detect active nodes and delivers a message to areas unvisited by agents. Therefore, agents should be confined into the areas of the network where messages will spread, in order also to enable the network graph used for delivery to grow dynamically as the agent migrates. A channel will be included in the message delivery only if an agent has passed through it and, therefore, a node will be included only if the agent has been there.

2.5 Delivery of multiple messages with multiple message source

The assumption that all messages originate in the same node is extremely limiting. In order to extend this algorithm to messages originating in any node, we run multiple instances of the same algorithm on the network, allowing concurrent execution. To the purpose of this description, n is the number of nodes in the system. The result is that:

- The state of a converging channel is represented by a vector with n elements where each node's state is registered. Before the channel is added to the active graph, the channel is considered on closed state. Once the channel is active, if no message has been received from a certain node, the state of the element in the vector, corresponding to that node, is set to open.
- Each message is processed respecting the state of the message associated to the node where it has originated.
- Nodes can deliver up to n messages simultaneously, i.e. no more than one for each node. As before, if a second message comes from the same node, it is put in the buffer until the preceding message ends its process.
- An agent carries always with it an array containing, for each message source, the identifier of the last message received. Furthermore, when an agent goes out through a new channel, it carries another vector containing, for each source message, the identifier of the last message processed by the source of the new channel exactly before the agent migrates.

When an agent reaches a node, it is held there until, for each message source, the identifier of the last message received is greater than the corresponding holding value (if present) of the channel through which the agent has arrived.

In order to enable a node to generate a message, we must ensure that the graph is connected. In order to do that, all connections are turned into bi-directional ones.

It must be considered again that the agents held could eventually be released; in which case the process can be executed by analysing the messages each node broadcasts. Let us assume that message i is the one with the lowest identifier among those that from any node have not been delivered to all nodes. There must be a route from each node where i has not arrived, and each node on its route is blocked until i arrives. Thanks to network graph connectivity, i will spread to every node through each channel and will complete delivery in the system. No node will put i in the buffer, since it represents the minimum message identifier one could expect. When *i* has completed its delivery, next message will be the new minimum and will be processed likewise. Since message buffering is made with respect to individual source, each node's messages keep processes independent.

In order to hold agents, nodes coordination is needed. The value *j* for each node where the channel is held, i.e. *holding*(*j*), is fixed when the agent arrives for the first time. Since messages will be surely processed, we have ensured that *j* is processed and the held agent released.

2.6 Implementation problems

The basic assumption for the correct functioning of the above-described mechanism is that communication channels are of the FIFO type. Each part of information travelling in the channel – such as messages, agents, as well as any combination of the two – must keep this feature. This is not necessarily a mobile agent platform requirement. Operations that need the message or agent delivery on different data flows are usually mapped through sockets or other high-level devices as remote method call. When such operations insist on the same destination, FIFO quality cannot be preserved, since information previously sent through a flow can be received after another one sent through a second flow, depending on runtime support architecture. FIFO feature can be nevertheless directly implemented in a mobile agent server, by associating a queue containing messages and agents to be broadcast to a remote server. Thus, FIFO feature is structurally applied to the server architecture, though this may need important changes in the case of an already existing platform.

The mechanism assumes that runtime support keeps the network graph state and the way messages are exchanged. Each server must keep an identifier vector for active exit and entry channels, and a vector containing buffered messages for each channel.

3 Reliable communication by means of mobile groups

A basic goal in mobile agent systems development is the way through which reliability of agent-based applications is supported. One of the requisites for the reliability of such applications is the need to reliably coordinate the actions of a group of mobile agents members. Agents belonging to a group, for example should be able to send messages to all the members of the group, and to react in case of any member's failure.

One of the approaches to support an execution consisting of a set of processes in distributed systems is the *process group*. In a process group, processes communicate with each other by exchanging multiple messages forwarded to the whole group. To preserve consistency of the group members' state, communication protocols ensure some properties, such as atomic delivery (a message can be delivered by all processes, or by none), a message request (partial or total), and a kind of virtual synchrony in which the changes on members of the group (caused by events like collisions and process unions) are constantly ordered respecting the message delivery. In such an environment, processes able to work detect the same sequence of events, though, in reality, the latter may occur in an arbitrary order.

In traditional group systems, however, processes are essentially *static*. In order to ensure applications based on a group of agents similar to those present in traditional ones, we will introduce the concept of *mobile group*. A mobile group is an extension of the traditional concept of process group that can directly support migrant processes as if they were group members. Through mobile groups, a migrating process can change its distributed environment location *while belonging to a group*. Mobile groups ensure also message delivery and a sort of virtual synchrony. Such guarantees are given *notwithstanding its members' mobility*. Besides, mobile groups make process mobility not only visible to the group, but also constantly coordinated with the other actions of the group. An implementation of mobile groups through the use of conventional group systems is not satisfactory, because process mobility would be hidden by the chronology of the actions of the group.

Thanks to the guarantees they give, mobile groups are a suitable abstraction for the development of applications based on reliable agents.

3.1 System model

Let us think of a distributed system as a set of mobile and static *communication processes, locations and channels*. A location represents the logical place in a distributed environment where processes are executed. When a mobile process migrates, it moves from one location to another. Processes set in different locations communicate through message exchange in communication channels.

On the one hand, we use $L = \{l_1, l_2, \ldots, l_n\}$ to show the set of all possible locations; on the other hand, we identify the set of all possible processes with P. A mobile group is identified by the set of processes $g = \{p_1, p_2, \ldots, p_m\}$ where $g \subseteq P$. We can define four basic operations in a mobile group:

- *Join* (g, p): used by p process, when it wants to join g group;
- *Leave* (g, p): used by p process, when it wants to leave g group;
- *Move* (g, p, l): used when a mobile process p wants to move from the present location to l location;
- *Send* (g, p, m): used by p process when it wants to send a multiple message to the members of g group.

We make a distinction between a message *received* in a location through communication channel and one *delivered* to p process. Delivery of a message to the process can be deferred to satisfy the synchronization demands of the system.

A p process is able to install views. In mobile groups a view $v(g) = \{(p_1, l_1), (p_2, l_2), \ldots, (p_n, l_n)\}$, with $p_i \in g$, $l_i \in$ L, $\forall\, i$ is a mapping between group g and locations. A view represents the set of a group of members considered mutually operative at a certain moment of the group's existence and indicates the locations where members are (a couple (p, l) in a view indicates that process p is situated in

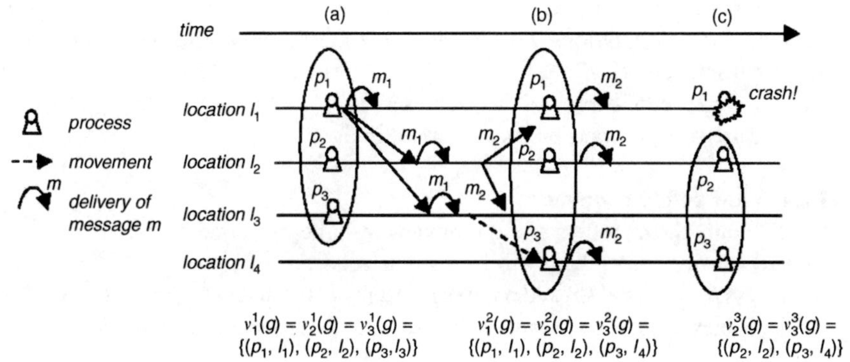

Figure 3: An example of mobile groups.

location l). Such set can dynamically change when process crashes occur or when processes deliberately leave a group, join it or move from a location to another. A new view is installed by the (operative) member processes every time a change in the group's view occurs. Each view installed by a process is associated to a number sequentially increasing with the installations of the group's view. In a group $g = \{p_1, p_2, \ldots, p_n\}$, $v_i^j(g)$ indicates the view number j installed by the process p_i.

Fig. 3 illustrates a group initially composed by three member processes, p_1, p_2 *and* p_3. These processes install in turn the views $v_1^1(g)$, $v_2^1(g)$ and $v_3^1(g)$ (fig. 3(a)). Such views are identical and indicate that processes p_1, p_2 and p_3 consider another process as operative and each process knows the present position of the other(s) $v_1^1(g) = v_2^1(g) = v_3^1(g) = \{(p_1, l_1), (p_2, l_2), (p_3, l_3)\}$.

Later, process p_3 moves to location l_4 (the movement is shown by dotted line in fig. 3(b)). Each process installs a new view, confirming the fact that process p_3 is now in a new location, the l_4 one views $v_1^2(g) = v_2^2(g) = v_3^2(g) = \{(p_1, l_1), (p_2, l_2), (p_3, l_3)\}$. After a certain lapse of time, in location *ll*, where *pl* is, a crash happens (fig. 3(c)). Thus, a new view will be installed by processes p_2 and p_3, confirming the fact that process *pl* has been removed from the group views $v_2^3(g) = v_3^3(g) = \{(p_2, l_2), (p_3, l_4)\}$.

We are certain of the reliability of communication channels because message delivery is sequential (FIFO). We assume that message broadcast and process time cannot be precisely estimated (because it could be an asynchronous system) and that the process fails its function and stops only when a crash occurs. Besides, we take into consideration that most group processes do not crash.

3.2 Properties

The consistence of process-installed views in a group is ensured by a *membership protocol.* The main aim of this protocol is to make sure that each member of the group installs the same view sequence. This would give the members of the group a mutually consistent group view as regards process crashes as well as *join*, *leave* and *move* operations. Furthermore, the messages sent to the group would be synchronized with all these events. Thus, group operational processes "would see" the same set of events as the group's.

Here are some properties (as illustrated in fig. 3):

3.2.1 View safety properties

Validity: only the members of a group view install corresponding views;

Single View Sequence: there will be a single sequence of views for each group. If process p_i installs $v_i^k(g)$ and process p_j installs $v_j^k(g)$, then $v_i^k(g) = v_j^k(g), \forall\, i, j, k$. Fig. 3 shows that the views with the same number (upper index) are the same.

3.2.2 Message delivery properties

Atomicity: any couple of g member processes installing two consecutive views deliver the same set of messages. Fig. 3 shows that messages m_1 and m_2 are delivered by all processes of the same view (m_1 in view v_i^1 and m_2 in view v_i^2). Message m_2, for example could not be delivered to view v_1^1 by process p_1, and by another process (p_2) to view v_2^2.

Existence: If a process p_i sends a message m to view r, then in the hypothesis that it carries on its function as a member of g, it will deliver m to another view $v_i^{r'}$, where $r' \geq r$. A message sent to a view might be delivered to a future view. This is shown in fig. 3, where message m_2 has been sent to view v_2^1, but delivered to views v_i^2.

Such properties enable processes to apply a certain virtual synchrony, i.e. processes in a group will install consistent views and will see the same set of messages in consecutive views.

3.3 A typical case

In this section, we are going to describe the quite common situation of an application making use of mobile agents to automate a process aimed at booking the user a seat on a plane. Among other requirements (price, class, etc.), the user asks to compare the prices of a certain number of airline companies – three at least – before confirming booking.

In general terms, such a situation can be implemented as follows: three agents are first of all created; each of them will determine the company's fare satisfying the customer's needs. Having an *a priori* list of airline companies to visit, each agent chooses one and moves to the site to check ticket's fare. If the result is not successful (e.g. there are no available seats in that company's flights), the agent migrates to another company's site among those included in the list and repeats the process until it succeeds.

Agents' actions must be coordinated not only to compare offers, but also to make the group react consistently on failures and exceptions. For example if one of the agents fails, another one must be created in order to complete the set of three agents searching for fare offers. Besides, when an agent wants to visit a new company, it should choose one that has not been already visited by other agents of the group.

Mobile groups can support coordination among agents and could be used as shown hereafter. The group of agents used in this context form a mobile group. Each agent represents one of the processes of the group. Every agent will have a consistent view of the group configuration thanks to the mobile group support. If an agent fails, the other agents know that and will act consequently. Every new view can have a group *coordinator*, defined through the application of a certain rule on a set of processes in that view (e.g. a coordinator is the process having the highest identifier in the group). A coordinator will be the agent that, for example creates new agents to carry out the task of those that had failed. Furthermore, since mobile groups give a consistent view of process locations, when an agent decides to visit a new company, it will not move to a location already visited by another agent of the group.

Views reveal, in particular, the agent's movements, possible agent's crashes and their joining (or leaving) the group. When an agent installs a new view, it can react in the following way: if it sees that a new agent has joined the group, it forwards its state (the companies visited before and the known fares) in order to inform new members of the group of former actions; if the view indicates that some member of the group has moved (an agent already present in the group has changed its view location), it adds the new location (drawn from the view) to the other locations visited; if the view reveals that the agent itself has a new location in it, that means that it has moved and, therefore, interacts with the local server and tries to obtain the ticket fare. Furthermore, the agent will apply a previously arranged function on the identifiers of agents in the view in order to be its coordinator.

Group system, besides, informs a process when it is not anymore a member of the group. In such a case, the process (or agent) exits.

When an agent detects an offer, it informs the user. When the application confirms that the fare has been received, the agent informs the group by sending it a message. When a member of the group receives at least three messages of this kind by other members, the application has got at least three ticket prices. So the agent leaves.

View coordinator checks regularly if there are fewer members in the group, verifying their number in current view. If need be, it creates new agents in order to complete the group.

4 Coordination through communication

In multi-agent applications, a distinctive key of software agents is social behaviour. Agents exchange information in order to achieve the application global goal. Research proposes two approaches as far as agents interaction is concerned:

- the adoption of ACL based on direct interaction between agents;
- the use of coordination infrastructures offering support to indirect interaction based on shared spaces (e.g. blackboard, tuple space, etc.) through which agents interact giving and receiving information.

Each approach has several advantages, but none of them give a complete framework satisfying all the generic multi-agent application requirements. Since agents are software entities able to act in a nearly human way, every agent needs to manage well-structured conversations directly with another agent. In other cases, agents need to share information, and adopt an indirect model to manage knowledge spreading. In order to provide an infrastructure for agent interaction coordination as much complete as possible, we have to combine both interactions, the direct one and the indirect one, in a single framework, so as to offer an almost human model of agent interaction based on the exchange of communication methods.

We must not forget that communication among agents occurs in the context of a well-defined multi-agent application possessing also a suitable set of interaction protocols planned for the application.

Such protocols establish the rules of every agent communication and can be used to develop laws apt to rule the infrastructure coordination managing agents interaction in this application. In order to obtain it, we provide a coordination model for both static and mobile agents, based on abstract structures called communicators, representing entities able to manage the agents dialogue made through ACL method.

A communicator is totally programmable, i.e. it is possible to specify which messages can be exchanged and how to manage them. Since every exchanged message passes through a communicator, the latter makes a syntactic and semantic routing, enabling the message exchange and forwarding according to the previously established coordination laws. The interaction classes establish them, i.e. a specification of interaction allowed in terms of the agent's role, of the kind of discussion acts, and of the management procedure.

4.1 Abstract models of interaction

When we plan a generally valid coordination framework, we must analyse the main characteristics of agent interaction and provide an abstract model for the latter.

The first step to plan a multi-agent application is to define role model and interaction model, whence interaction protocols and agents' behaviour in the context of defined protocols derive. Afterwards we can derive the first property of agent interaction:

Any interaction between two or more agents sharing the same multi-agent application may occur if (and only if) that application's role and interaction models allow it.

Such a model is not complete to represent an almost human behaviour. Human interaction not only happens one to one, but also one to many, and receivers can be reached explicitly or implicitly; for example in a room where many people are, if a single person talks explicitly to his/her friends, conversation could be (implicitly) followed by others. That means that we must take into consideration both the agents to whom a message is explicitly sent and the agents that happen to "listen" to a message, even if it is not directed at them. These points can be modelled by introducing the set of recipients and the goal. They define what is called real and virtual recipients. Real recipients are the agents to whom the message is explicitly addressed; the virtual ones are all the agents that can share the message.

Virtual recipients cannot be under the sender agent's control: it (like a human being) can specify which message to send, the agent whom it is speaking to, etc., but the aim of the interaction is determined by the application context where the interaction takes place. That is why we model interactions by specifying the association between the message and its virtual recipient. This association depends on the application requirements and, consequently, on the interaction semantics in the context of that application. In order to allow a flexible specification in such associations, we group all interactions having the same behaviour in interaction classes generalizing every interaction field.

4.2 Communicators

In order to provide a coordination infrastructure following the interaction model described above, we introduce coordination entities called *communicators*.

4.2.1 Simple communicators, *SK*

They are the entities represented by the triplet $SK := <\overline{I}, \{m\}, \{\alpha\}>$: $\overline{I}$ is an interaction class as defined before, $\{m\}$ a set of ACL messages temporarily stored in *SK* and $\{\alpha\}$ the set of agents that joined *SK*. A set of primitives provides the operations that can be done on *SK*, allowing the agent to join or leave *SK* and read or write the message. Communication between agents takes place through the exchange of messages stored in *SK*, whose task is to control that the message belongs to the interaction class $\overline{I}$ and then behave as a *semantic router* forwarding the message to the agent/s according to the interaction class definition. The basic *SK* primitives an agent can use are $join_{SK}$, $leave_{SK}$, put_{SK} and get_{SK}. The first one must interact with an agent before executing any communicative act (send or receive a message); $join_{SK}$ adds an agent a to the set of agents $\{\alpha\}$ only if a's role is defined (as sender or recipient) in interaction class $\overline{I}$. Later, the agent can send or get back a message by using put_{SK}, and get_{SK} primitives: the first one puts the message in the set of messages $\{m\}$ only if the agent belongs to the group $\{\alpha\}$ and the message to $\overline{I}$. Get_{SK}, instead, gets back a message m' from $\{m\}$, only if the agent a belongs to the set $\{\alpha\}$, a's role is defined as real recipient (the message is taken from $\{m\}$) or as virtual recipient (the message is *read but not deleted* by $\{m\}$) in $\overline{I}$. To add flexibility to our model, we also introduce p_get_{SK}, which acts like get_{SK} but returns a message together with a certain predicate. Finally, an agent can leave the communicator using the primitive $leave_{SK}$.

4.2.2 Composite communicators, *CK*

They are defined as entities composed by a set of *SK*, $CK := \{SK_1, SK_2, ..., SK_n\}$. A *CK* provides the same primitives as *SK*, but they behave in a slightly different way; in particular, $join_{CK}$ adds the agent to each SK_i, put_{CK} inserts the message in each SK_i which makes the put_{Sk_i} predicate come true; get_{CK} (p_get_{CK}) recovers a message from a SK_i which makes the get_{SK_i} ($p\_get_{SK_i}$) predicate come true; finally, $leave_{SK}$ removes the agents from each SK_i group.

Such a communicator is able to structure the right relations and therefore to provide facilities for a given application coordination defining the interaction classes implemented in *CK*. That is to say that, once the multi-agent application has been planned and interaction classes defined, we can carry out the communicator able to support a given application message exchange. Each agent, before taking part in the application, must join its communicator; it can later communicate with other agents sending and receiving messages through such communicator.

The model shown above is able to carry out both *direct* and *indirect* communication. The former is obtained by stating precisely the addressee agent(s); the latter is obtained by stating precisely *all* real and virtual recipients: according to the composition of the sent message, it will be put in the corresponding *SK* and later read by the agent belonging to that *SK*'s set of *real* and *virtual* recipients.

4.3 ACL

Dealing with communication, we cannot neglect what directly interests interlocutors, i.e. language. It can be defined as an ensemble of signs, symbols and objects through which communication can take place.

Virtual knowledge base is an expression indicating an ensemble of declarative information and a deduction mechanism enabling the agent to answer questions and even take autonomous decisions. Within this context, interoperability is often interpreted as a problem concerning the agent's ability to share its own information and knowledge with others. Problems concerning the store of knowledge carried by agents nevertheless arise, which can be outlined in three major points:

- How to pass from a language to another?
- How to ensure that the meaning of objects and concepts is the same to different agents?
- How agents exchange (communicate) this part of knowledge?

ACL enables us to solve the problem of interoperability.

An ACL can be thought of as a set of messages of any kind with a particular meaning. A communication language has nothing to do with an exchange of expressions, but rather it refers to their content and their consequent attitude. Therefore, there is a clear difference between various kinds of messages: request, assertion, interest for a particular content, etc.

From the engineering point of view, an agent communication language can be seen as a protocol of messages with two basic differences:

- It describes application and actions at a higher abstraction level.
- It offers a wider range of messages.

An ACL, then, can be seen as a structure offering assistance in the hard problem of application interoperability, even if that is not the only problem to be solved: dangers resulting from wireless and wired environment heterogeneity must also be taken into account.

An ACL offers three kinds of advantages:

- It supports interoperability among static and mobile agents as well as among mobile agents planned for different platforms.
- ACL's declarative nature enables us to have at our disposal many services that make information exchange easier.
- The higher level of abstraction at which ACL operates enables us to interface also multiple paradigms.

5 Knowledge sharing effort (KSE)

Knowledge sharing effort (KSE) was introduced in 1990 by DARPA. Dozens of researchers participated whose aim was to develop software techniques, methodologies and instruments for information sharing and its consequent re-use. KSE's basic concept was that sharing needed communication, and hence a common language. Research focused on the definition of a common language; in the model developed, software systems are seen as "knowledge virtual bases" exchanging propositions by the use of a language expressing various and complex attitudes. Agents, therefore, were not an integral part of KSE at the beginning.

The first thing to do is to convert languages into the same language family. The second is to ensure that the semantic content of signs is kept in applications, in other words that the same concept, object or entity has a uniform meaning throughout applications, although it can have different names in different languages. The technical term for this knowledge background is *ontology*; it is, more formally, a particular conceptualisation of a group of objects, concepts and other entities that are the expression of information and relations. Ontology consists of terms, their definition and axioms. The third and last thing to do is to inform communication among agents; it does not concern though the relation between bits or bytes: it is rather the expression of an exchange of information with sometime complex content. Mobile agents need to ask other agents to inform them, to find other agents giving assistance, to display values and objects, to ask for a particular service, etc.

An ACL is seen just as a collection of propositions expressing what the agent would like to say. KIF, a particular kind of logical language, was proposed within KSE as a standard used to describe a computerized system's inside, i.e. databases, intelligent agents and so on; furthermore, it was specifically programmed to produce such an useful thing as an "interlanguage", i.e. a language able to translate from/to various languages. Language description includes both syntax and semantics. *Ontolanguage*, a variety of support tools, was KSE's solution to development problems: Stanford research laboratories developed a set of tools and services to obtain the approval of groups geographically spread over a common ontology. Such tools are created around "*ontology*", a language planned to describe ontologies; it makes use of the web putting the user in a position to publish or create new ones on an ontology server.

5.1 KQML

KQML is a high-level communication language that can be seen as a protocol for the exchange of information with independent content, syntax and ontology. It is, therefore, independent of transport mechanisms (TCP/IP, SMTP, etc.), of language content (KIF, SQL, etc.) and of ontology. We can classify it in three layers:

CONTENT
MESSAGE
COMMUNICATION

The first layer carries the current message content ignoring its extension (except for determining its end) and should also recognize its language. KQML can transmit any language, included ASCII code strings and expressions that make use of binary code.

The second layer is the kernel of KQML. It is used to decode a message that an application would like to forward to another and determines the various kinds of interaction among agents. Primary function of this layer is to identify the protocol used to save the message and provide a way to chat or the representation of what the sender means (question, statement, etc.).

Should the content not be clear to KQML, it is equipped with optional services apt to describe language content, to adopt ontology and to search the subject in question; such services enable KQML a thorough analysis of the message.

The third layer interprets a set of message services describing the lowest level of communication parameters.

KQML syntax is based on Lisp's familiar expressions.

Since language is relatively simple, effective syntax is insignificant because it could be updated.

Here is the structure of a KQML message:

(*ask-one* — it expresses the message content that in such case is a question (ask);

: sender... — it contains the name of the sender agent

: content(...)	it carries the question expression (...);
: receiver...	it contains the name of the receiver agent ...;
: reply... ...	it informs us that the message is a reply ...;
: language...	it informs us on the language used by the agent to write the message;
: ontology(...)	it informs us on the ontology of the message.

One of the criteria leading to the development of KQML was to produce a language that could support a wide range of architectures and agents. KQML also introduces a special class of agents called *communication facilitators*. A facilitator is an agent that, though keeping a service register, manages messages providing for service mediation and translation.

5.2 FIPA

Though KSE introduced the main points concerning research and approach to the problem of agent interoperability, it did not offer disciplined ways to model development. This gap was filled in 1996 by the Foundation for Intelligent Physical Agents (FIPA), a non-profit association aiming at promoting the success of emerging agent-based applications. FIPA's merit is undoubtedly the maximization of agent-based systems interoperability. FIPA operates through international cooperation of companies and universities working in this research field. FIPA operates by assigning tasks to technical committees (TCs) that have to carry out and maintain them.

A TC is assigned a task and it must return a detailed description to an ACL. FIPA ACL, like KQML, is based on a theory according to which messages are actions or communicative acts, every act is described both in narrative form and through logic-based semantic form. Syntax is almost the same as KQML's, except for some different primitives' names. KQML's approach to separate outer from inner language is therefore unchanged; the former informs us on content, the latter on the interlocutors' intentions and desires. A detailed FIPA ACL document requires, like KQML, that language not to make any use of any particular language content.

5.3 ORB and CORBA

With object request broker (ORB) we mean an object message exchange agent. ORB represents the structure on which distributed objects communication is based.

OMG, object management group, defined a common architecture for ORBs enabling the birth of the specific CORBA (common object request broker architecture).

In order to link to an ORB, an object needs an extra software layer called stub on the client side and skeleton on the server side. CORBA is an open standard, not linked to any particular producer, defining the architecture for the creation of object-oriented distributed applications. CORBA is very powerful and is not linked to any particular development language, thus allowing interfacing among heterogeneous systems. Independence from the language used in CORBA is obtained thanks to IDL (interface description language), a language defining the object's actions through a mechanism independent from the programming language of object methods implementation.

5.4 RMI

Remote method invocation (RMI) is a powerful and simple set of APIs, allowing the development of network distributed applications. RMI has its own ORB not responding to specific CORBAs. That means that objects using RMI's ORB could only talk with objects using the same ORB.

Javasoft ownership protocol on which RMI is based is called JRMP (Java remote method protocol). Furthermore, the "contract" between client and server, which in CORBA is defined by IDL module, in RMI is "drawn up" by Java interface. Consequently, in order to use RMI, one does not need to know IDL, Java being the only development language.

RMI is basically a simple though closed technology, both client and server must be written exclusively in Java.

RMI use is advisable when a full-java distributed object structure must be implemented. If, instead, distributed applications with parts already made with different languages must be carried out, the ideal choice would be the use of CORBA.

5.5 RMI-IIOP

RMI-IIOP was developed by Sun and IBM in collaboration with OMG to develop a new RMI version able to communicate not only with Sun JRMP, but also with Internet inter-ORB protocol (IIOP).

CORBA specification introduces a general inter-ORB protocol (GIOP) defining messages format and data representation for all distributed object communications on the same or on different ORBs, in order to enable interoperability among different ORBs.

A particular implementation of GIOP, precisely called IIOP, uses TCP/IP protocol to implement the transport layer for messages between remote objects.

IIOP is therefore a kind of "glue" relying on TCP/IP among the objects distributed on different ORBs.

The new version of RMI (RMI-IIOP) uses IIOP, representing *de facto* the first meeting point of until now mutually incompatible RMI and CORBA. RMI-IIOP is fully compatible with RMI-JRMP and enables communication (with some limitations) with CORBA applications.

6 Synchronization

There are two main kinds of synchronization: *synchronous* and *asynchronous* [1, 2, 3, 4].

In *synchronous* communication, agents need to synchronize before transferring data, so the sender agent stops on the "*send*" primitive and the receiver agent on the "*receiver*" one, until the latter receives communication from the sender agent. A direct consequence of this kind of communication is that when the sender agent completes the execution of the "*send*" primitive, it knows that the receiver agent has received the information forwarded. In *asynchronous* communication, on the contrary, the sender agent does not stop on the "*send*" primitive, so the sender agent does not know the forwarding outcome. In such case the sender agent can decide to stop or not on the "*receive*" primitive, if it does it must stay there until the end of communication.

7 Location

Location is another aspect to be kept into account in order to know whether agents' interactions are local or remote. When agents meet in a place to interact, communication is local; when they communicate from different places, communication is remote.

For this reason two communication mechanisms can be distinguished [10]:

- Location-dependent;
- Location-independent;

In *location-dependent* communication two or more agents communicate with another agent in a particular location, called *exchange post* in distributed system. In order to communicate, it is necessary that the sender agent first and the receiver agent later, or both simultaneously, visit that specific location. In *location-independent* communication, on the contrary, two or more mobile agents communicate without taking into consideration a particular network location.

In the following sections *location-dependent* and *location-independent* communication will be analysed in detail.

7.1 Location-dependent communication

Location-dependent communication mechanism involves a sender agent S on the node S_l, a receiver agent R on the node R_l, information M that must be communicated and a node N in the network.

Communication is executed in two steps:

- When S executes the primitive "send", information M is stored in node N;
- When R executes the primitive "receive", information M on node N is restored and delivered to receiver agent R.

A sender agent can locally execute the primitive "*send*" while it is on node N ($N = S_l$), or remotely while it is on a different node than N ($N \neq S_l$). Likewise, a receiver agent can execute locally the primitive "*receive*" while it is on the node N ($N = R_l$), or remotely while it is on a different node than N ($N \neq R_l$).

According to what we have just said, *location-dependent* communication can be divided in three types:

- Local sender and receiver (LSLR);
- Local sender and remote receiver (LSRR);
- Remote sender and local receiver (RSLR);

In the *local sender and receiver* kind of *location-dependent* communication, shown in fig. 4, both sender and receiver agents in turn execute the primitives "*send*" and "*receive*" while they are on the node N ($N = S_l = R_l$).

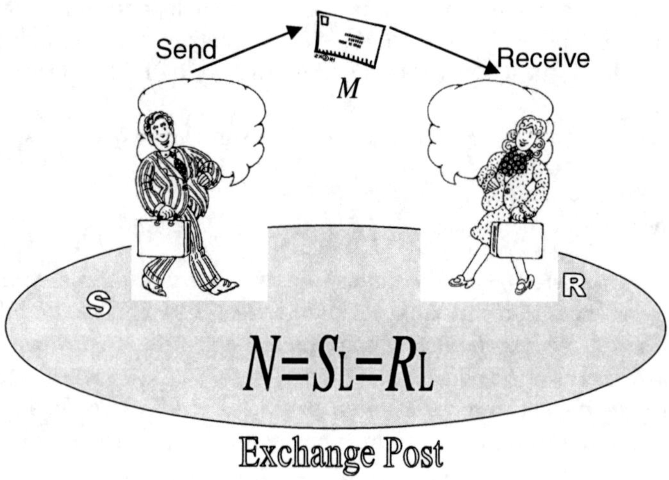

Figure 4: Local sender and receiver (LSLR).

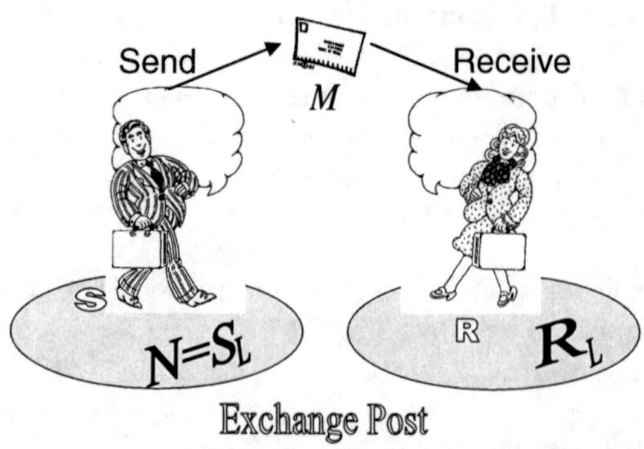

Figure 5: Local sender and remote receiver (LSRR).

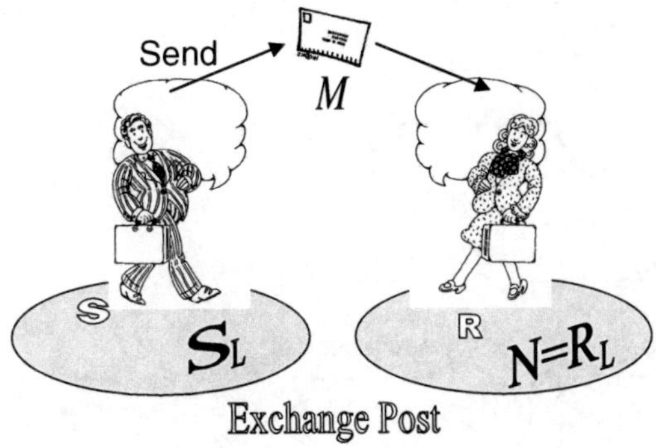

Figure 6: Remote sender and local receiver (RSLR).

In the *local sender and remote receiver* case, shown in fig. 5, the sender agent executes the primitive "*send*" while it is on the node *N*, while the receiver agent executes the primitive "*receive*" while it is on a different node than *N* ($N = S_l$, $N \neq R_l$).

Finally, in the *remote sender and local receiver* case, shown in fig. 6, the sender agent executes the primitive "*send*" while it is on a different node than *N*, while the receiver agent executes the primitive "*receive*" while it is on the node *N* ($N \neq S_l$, $N = R_l$).

There exists also a fourth possibility, *remote sender and remote receiver*, but this is a particular case of *location-independent* communication that will be analysed in the following section.

7.2 Location-independent communication

A specific location visited by both sender and receiver agents to communicate does not exist in location-independent communication. Since agents are mobile, the sender agent possibly does not know the exact location where the receiver agent is or will go to. A problem thus rises, called *location problem*, of how to make mobile agents communicate. In order to solve this problem, we can resort to message-passing communication models.

8 Models

Eight models of communication have been obtained from the combination of communication and synchronization mechanisms, as shown in fig. 7.

The eight models are divided into three classes:

Class 0 includes models 0 and 1. Models in this class are the result of the combination of LSLR communication with one of the two synchronization techniques;

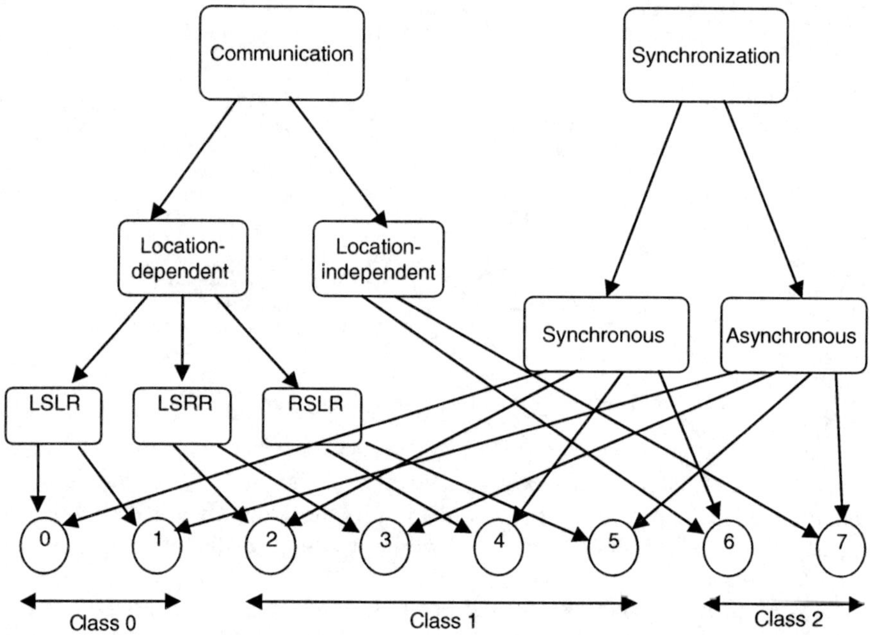

Figure 7: Models.

Class 1 includes models 2, 3, 4 and 5. Models in this class are the result of the combination of LSRR/RSLR communication with one of the two synchronization techniques;

Class 2 includes models 6 and 7. Models in this class are the result of the combination of *location-independent* communication with one of the two synchronization techniques.

9 Message-passing

As to *Class* 2 models, since agents are mobile and a specific location where mobile agents can communicate does not exist, the *location problem* rises. It consists in finding the receiver agent's location. This problem can be solved by using some communication models based on *message-passing* [12], i.e. on communication between two or more mobile agents through message exchange. In this section some kinds of models using *message-passing* [11] to make mobile agents communicate are introduced.

The five communication models taken into consideration are:

- Home-Proxy;
- Follower-Proxy;

- E-mail;
- Blackboard;
- Broadcast.

Each model describes an approach that enables an agent (*static* or *mobile*) to send a message to a mobile agent.

9.1 Home-Proxy

In the *Home-Proxy* model, e.g. "*Aglets*" [5], both sender and receiver agents operate under the following context:

- The sender agent knows the receiver agent's name;
- The receiver agent has a home-place (main location);
- The sender agent has access to a lookup service allowing him to know the address of the agent's home-place where it intends to send the message;
- The receiver agent is a mobile one.

If the agent uses a *Home-Proxy* model, as shown in fig. 8, it must make use of a *lookup* service to know the address of the receiver agent's *home-place* before sending the message.

Later the sender agent sends the message to the receiver agent's *home-place*. If the latter is in its own *home-place*, the message is immediately delivered. If the receiver agent, being mobile, is in another location, it is its responsibility to inform its own *home-place* about where it is each time it changes location. When the *home-place* receives a message, it will be delivered to the receiver agent's location. In such model, the *home-place* must have a database to store the address of the location that the agent is using temporarily as its *home-place*. Depending on implementation, database can be updated before the receiver agent leaves a particular location to move to another one, or after it arrives. In extreme case, the *home-place* could store the addresses of all mobile agents in the system, in which case *lookup* service is not necessary.

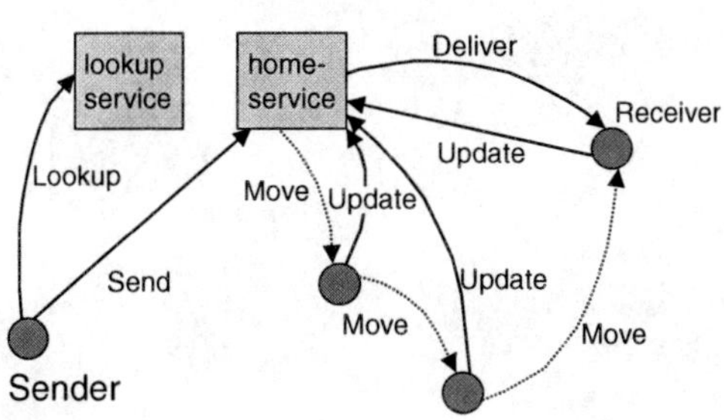

Figure 8: Home-Proxy communication model.

9.2 Follower-Proxy

In the *Follower-Proxy* model (fig. 9), e.g. "*Voyager*" [6], both sender and receiver agents operate under the following context:

- The sender agent knows the receiver agent's name;
- The receiver agent has a home-place (main location);
- The sender agent has access to a lookup service allowing it to know the address of the agent's home-place where it intends to send the message;
- The receiver agent is mobile.

At a first sight the Follower-Proxy model, shown in the figure below, is similar to the Home-Proxy one.

The sender agent's first step is to use the *lookup* service to find the address of the receiver agent's *home-place*, where it sends the message. If the receiver agent is in its own *home-place*, the message is delivered immediately. So far the *Follower-Proxy* model is the same as the *Home-Proxy* one. Differences come up when the receiver agent moves. In this model, the receiver agent has the responsibility to inform the last place visited about its current location. When the receiver agent moves from its *home-place*, it will forward the message to the location where the agent has moved. If the agent has moved to another location, the message would be forwarded to that node, and so on. To deliver a message to an agent, the message must browse through a list of forward links, whose last one is the receiver agent's current location. In this model every place must have a database containing the addresses of the moving agents forwarders.

As in the *Home-Proxy* one, a database can be updated before the receiver agent leaves the place, as well as after it arrives. In extreme case, all locations could contain the addresses of all the agents in the system, in which case *lookup* service is not necessary.

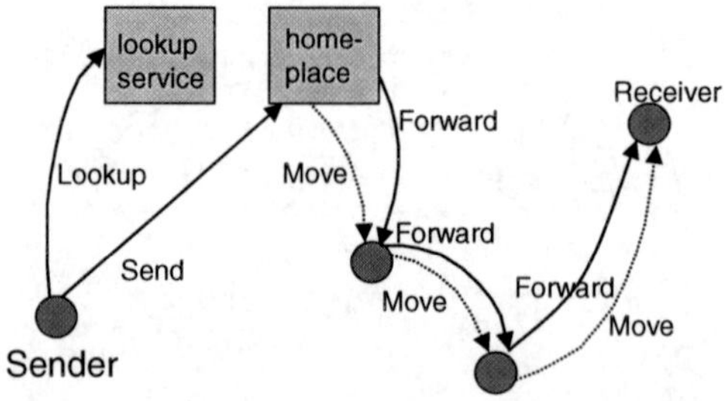

Figure 9: Follower-Proxy communication model.

9.3 E-mail

In the *E-mail* model (fig. 10), e.g. "*inbox*" in *ffMAIN* [7], both sender and receiver agents operate under the following context:

- The sender agent knows the receiver agent's name;
- The receiver agent has a home-place (main location);
- The sender agent has access to a lookup service allowing it to know the address of the agent's home-place where it intends to send the message;
- The receiver agent is mobile.

This model is similar to the *Home-Proxy* and *Follower-Proxy* ones. An agent using *E-mail* model, as shown in fig. 10, sends a message to another agent by using at first a *lookup* service in order to detect the receiver agent's *home-place*'s address. Then the sender agent sends the message to the receiver agent's *home-place.* This will be added to a list of messages of the receiver agent's stored in the *home-place* local memory. It is the receiver agent's responsibility to check the list of messages periodically to detect messages. In such model, each location must have a database to store messages for the agents having their main location there. In extreme case, a node could store the messages of all mobile agents in the system, in which case *lookup* service is not necessary.

9.4 Blackboard

In the *Blackboard* model (fig. 11), e.g. "*AMBIT*" [8], both sender and receiver agents operate under the following context:

- The sender agent does not know the receiver agent's name;
- The receiver agent can or cannot have a home-place;
- The receiver agent is mobile.

In the already seen three communication models, a sender agent has been assumed to know the receiver agent's name. Therefore, it finds an agent's location

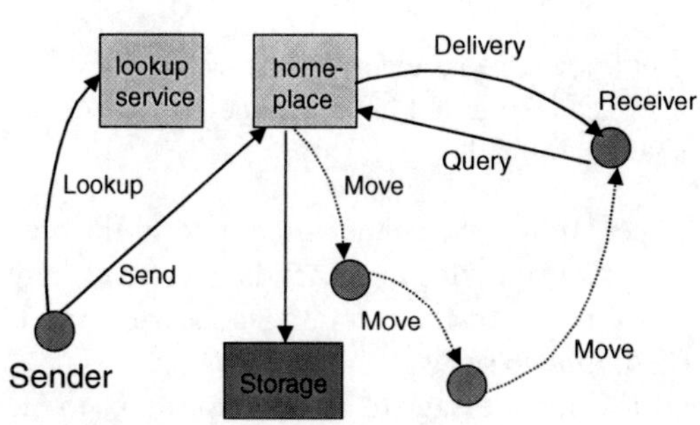

Figure 10: E-mail communication model.

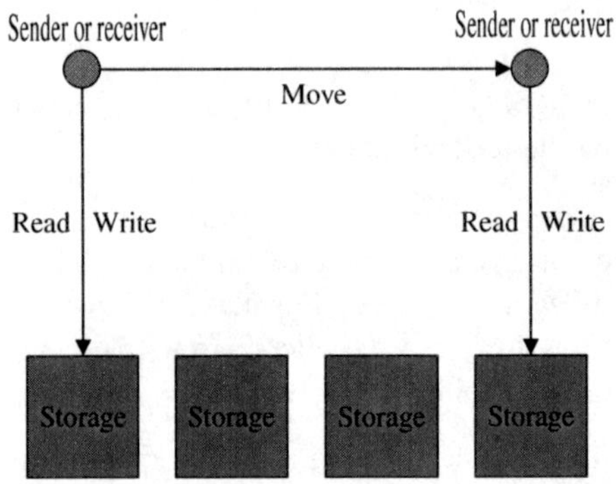

Figure 11: Blackboard communication model.

by using a *lookup* service. In *Blackboard* communication, on the contrary, as shown in the following figure, the receiver agent's name and location are unknown.

Communications will be anonymous. *Blackboard* communication implies available memories in every location, which every agent can use to store messages for other agents or to read them. As shown in the figure, an agent sends a message to another by writing on a node's currently not inhabited local memory, then it will move to another place, repeat the operations and so on. In order to read or deliver a message, a receiver agent must move to the place where it is stored and then read the local memory. In such model, every location must have a database and be able to deliver and manage messages. In extreme case, a place could store messages contained in every other location because agents store messages in all the locations they visit.

9.5 Broadcast

In *Broadcast* model [9] both sender and receiver agents operate under the following context:

- The sender agent knows the receiver agent's name;
- The receiver agent can or cannot have a home-place;
- The receiver agent is mobile.

In the models seen above, the sender agent knows the receiver agent's name and makes use of a *lookup* service to detect the location where the message is to be sent. In *Broadcast* model, on the contrary, as shown in fig. 12, the sender agent does not make use of *lookup* service.

It simply broadcasts the message to all other locations in the system. Location itself will deliver the message to its agent if it is currently in that location, otherwise the message will be stored for a certain time and delivered when the agent arrives.

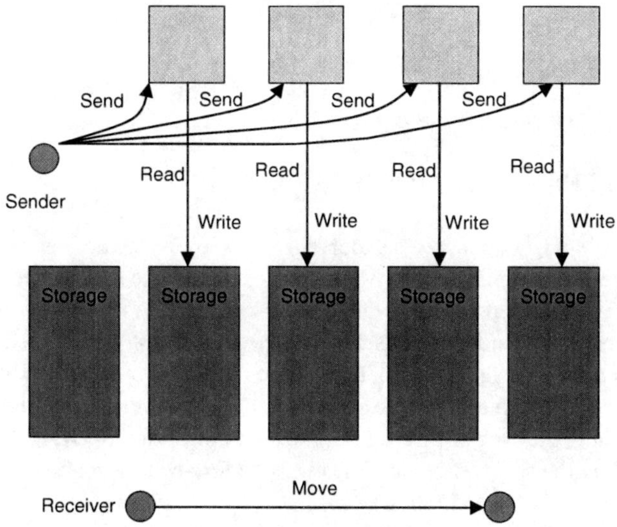

Figure 12: Broadcast communication model.

10 Cost estimation

When a communication model must be chosen and adopted in a mobile agent system, many factors must be considered, among which message and memory costs. The cost of the five models previously analysed will be examined in this section. For the sake of simplicity, let us consider messages and memory costs equal to 1, and let us suppose that the number of agents A coincides with the number of nodes N in the network.

10.1 Cost of Home-Proxy model

In *Home-Proxy* model, the sender agent, in order to deliver the message to the receiver agent, must send a message (cost C_l) to the *lookup* service (to ask for the receiver agent *home-place*'s address), which sends a reply message (cost C_r). Later, the sender agent sends the message (cost C_s) to the *home-place* that delivers the message (cost C_d) to the receiver agent in its current location. Cost of messages (cost C_u) sent by the receiver agent to its *home-place* multiplied by the M times it changes location, must be considered. The cost for sending the message is

$$\text{Cost of messages} = C_l + C_r + C_s + C_d + M(C_u) = M + 4$$

As to memory cost, the *lookup* service must store the *home-place* address (cost S_a) multiplied by the number of mobile agents (A) and their *home-place* and must store the address of the location where the receiver agent is (cost S_a).

$$\text{Cost of memory} = AS_a + S_a = NS_a + S_a = N + 1$$

10.2 Cost of Follower-Proxy model

In *Follower-Proxy* model, the sender agent, in order to deliver the message to the receiver agent, must send a message (cost C_l) to the *lookup* service (to ask for the address of the receiver agent's *home-place*) that sends a reply message (cost C_r). Later, the sender agent sends the message (cost C_s) to the *home-place* that delivers the message to the location where the receiver agent has moved (cost C_f), if the agent is not in that location, the message is delivered to the next location and so on. Thus, the cost of message C_f must be multiplied by M, i.e. by all locations visited by the receiver agent. The message forwarding cost is

$$\text{Cost of messages} = C_l + C_r + C_s + M(C_f) = M + 3$$

As to memory cost, *lookup* service must store the *home-place* address (cost S_a) multiplied by the number A of mobile agents and both the *home-place* and every visited location must store the address of the location where the receiver agent has moved to.

$$\text{Cost of memory} = AS_a + MS_a$$

At worst, the agent visits all nodes, so we have $M = N$. In such case, memory cost is

$$\text{Cost of memory} = AS_a + MS_a = NS_a + NS_a = 2N$$

10.3 Cost of E-mail model

In *E-mail* model, the sender agent, in order to deliver the message to the receiver agent, must send a message (cost C_l) to the *lookup* service (to ask for the address of the receiver agent's *home-place*) that sends a reply message (cost C_r). The receiver agent sends a request message (cost C_q) to its *home-place*, which delivers the message (cost C_d).

$$\text{Cost of messages} = C_l + C_r + C_s + C_d + C_q = 5$$

As to memory cost, the *lookup* service must store the *home-place* address (cost S_a) multiplied by the number A of mobile agents, and the *home-place* must store the message (cost S_m). Memory cost is

$$\text{Cost of memory} = AS_a + S_m = NS_a + S_m = N + 1$$

10.4 Cost of Blackboard model

In *Blackboard* model, the sender agent must write the message in the node's local memory (cost C_{wb}) in order to deliver it, then the receiver agent sends a message to be read (cost C_{rb}). Its cost therefore is

$$\text{Cost of messages} = C_{rb} + C_{wb} = 2$$

As to memory cost, the node must deliver the message (cost S_m).

$$\text{Cost of memory} = S_m = 1$$

10.5 Cost of Broadcast model

In *Broadcast* model, the sender agent must send a message to all the N nodes in the network (cost C_s), and the node where the receiver agent resides must deliver the message (cost C_d) to the receiver agent. A message forwarding cost is the following:

$$\text{Cost of messages} = NC_s + C_d = N + 1$$

As to memory cost, the N nodes in the network must store the message (cost S_m) for a certain time. Memory cost is

$$\text{Cost of memory} = NS_m = N$$

10.6 Model comparison

The *Message Cost* under mobility diagram in fig. 13 shows that *E-mail* and *Blackboard* models, in order to deliver a single message to the receiver agent, have always the same cost, being it independent from mobility, i.e. the number of nodes visited by the receiver agent. *Home-Proxy* and *Follower-Proxy* models, on the contrary, show an increasing trend as the number of nodes visited by the receiver agent increases. *Broadcast* model is not included since message delivery cost does not depend on mobility, but exclusively on the network extent (number of N nodes).

As to memory cost, the *Memory-Nodes Cost* diagram in fig. 14 shows that *Follower-Proxy* model has a higher cost than the other models, while *Blackboard* model has a lower cost. Cost of *Home-Proxy, E-mail* and *Broadcast* models have a similar trend, increasing as the nodes in the network increase.

After analysing the five *message-passing* communication models we realize that the choice of the model to be adopted in one's own mobile agent system depends on various factors, among which memory and forwarding costs. Moreover, choice depends also on factors such as the network extent, the mobile agents number, as well as security. Each model will be adapted to the exigencies of the mobile agent system one wants to create by combining such factors.

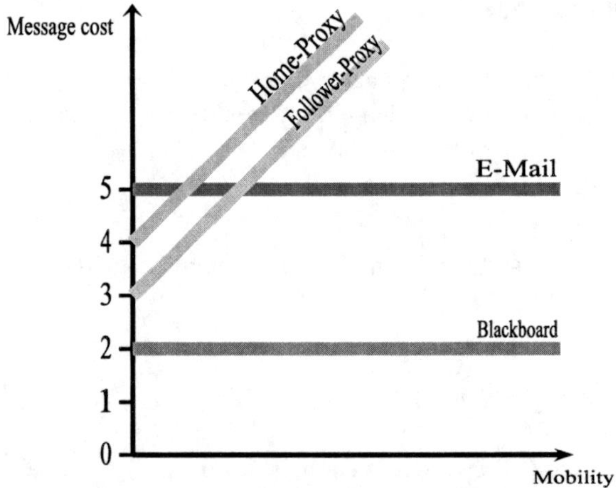

Figure 13: Message cost vs. mobility diagram.

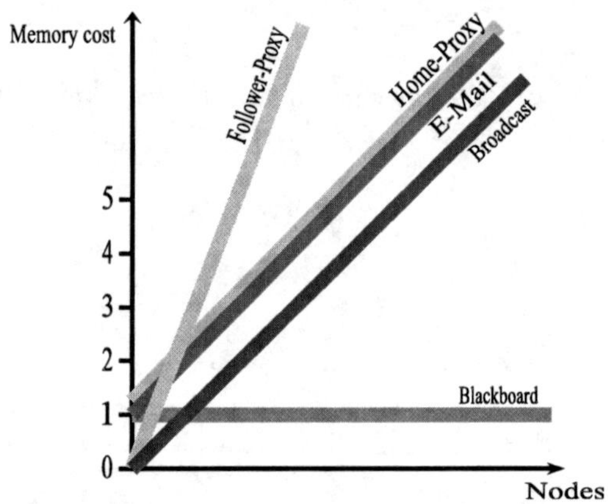

Figure 14: Memory cost under nodes diagram.

They can be thought of as levers indicating the compromise to be reached. The number of agents, the possibility of model failure, the complexity of model implementation, as well as security, among other factors, must be taken into consideration.

As a matter of fact, if we deal with this subject only from cost perspective, as far as messages are concerned *E-mail* and *Blackboard* models need a quite low number of exchange messages between two mobile agents to deliver a single one, while *Home-Proxy* and *Follower-Proxy* models need a high number of them. *Broadcast* model is not included in the figure because the number of messages used does not depend on the recipient agent's mobility, but exclusively on the network extent, *N*.

From storage perspective, *Blackboard* model needs the lowest amount of storage to support mobile agents messaging, while the *Follower-Proxy* one needs the highest one. *Home-Proxy, E-mail and Broadcast* models all need a similar storage amount that is lower than the *Follower-Proxy* one, but higher than the *Blackboard* one.

From this preliminary overview, it seems that mobile agents should use *Blackboard* model, because, to deliver a mobile agent message, it needs the lowest number of messages and the lowest amount of storage. There are though other circumstances to be taken into account before making a final decision.

When only few agents are in the final system, the use of a *lookup* service to find their *home-places* unnecessarily overloads the application, not only because of development and maintenance costs, but also for the execution performance of message sending. Furthermore, why provide a lookup service or *blackboard* model if the agents can autonomously keep trace of other agents' addresses? Broadcast messages seem more suitable if a few agents don't need to communicate very often. In the presence of a high number of agents, though, *broadcast* method floods the network with lots of unnecessary messages. Furthermore, if agents' *home-places* are not eventually distributed, location could be loaded by storage and by too many agents' message requests.

11 Fault causes

Faults can occur. Their implications must be taken into account when a system is planned rather than ignore them. In the previously presented model, the following faults can occur and compromise message delivery:

- Lookup service fails;
- A location fails;
- A message is lost;
- A site's storage fails.

If lookup service fails, *Home-Proxy, Follower-Proxy* and *E-mail* models cannot deliver the message. However, a lookup service can be replicated or planned to restart automatically, and the agent's information can be made persistent to minimize faults. Providing an auto-restarting and repeated lookup service means increasing the project complexity and, since application minimum code is required, development and maintenance costs increase.

If the failing location is the agent's *home-place*, *Home-Proxy, Follower-Proxy* and *E-mail* models cannot deliver messages. Besides, *Follower-Proxy* model will fail if any place visited by the agent fails. A common characteristic of all models is to fail if the receiver agent's location fails, since no messages can be delivered there. The same epilogue occurs when a message is lost in between two locations. Locations can be restarted; it does not mean, though, that location information is persistent. The consequences of adding a persistency layer to a site are the same as when additional components are added to the project: further development and maintenance costs.

In order to ensure a more reliable exchange of messages, some protocols adding additional costs of development and maintenance exist.

All models need some form of storage. *Home-Proxy* model requires a location to store the actual position of the agent during its migration. *Follower-Proxy* model uses all locations on the agent's route to store its next location. *E-mail* model must store the agent's messages in its *home-place*. *Blackboard* model stores messages in each location. Finally, *Broadcast* model needs locations to store the agent's messages temporarily. So, if storage fails, the model will fail too. Consequently, there is the need of a persistency layer in the software to restore memory when that kind of fault occurs.

12 Complexity

Blackboard model seems to be the simplest one. To deliver a message, an agent simply writes it on local device and another agent reads it. What actually happens is that message delivery complexity has moved inside the message exchange framework in the multi-agent system, towards the application developer. Being messaging anonymous now, developers must be sure that the right agents go to their locations to receive messages. If an agent cannot reach the location to receive the message, it is as if a message has not been delivered.

The model's device is itself a further complexity. How long must the message stay in the device? Who establishes when to remove the message? There cannot be messages staying forever, since space is not unlimited.

The simplest model would be then the *Broadcast* one. There are no lookup service, *home-place* location or forwarding to be managed. Here, though, its cost is the number of messages needed to deliver only a single one. If the number of agents in the system is considerable and they are sending many messages, overloading could not support application.

E-mail model seems to be a reasonable approach. The need for lookup service, local storage, persistency services, however, makes the model more complex to be projected, implemented and maintained. What we have just said applies also to *Home-Proxy* and *Follower-Proxy* models. The latter, however, has another complexity aspect to be dealt with. When should the information forwarded by the agent be removed? Who is responsible for it? Is it the agent or the location? What happens when an agent gets back to an already visited place?

The benefit of additional complexity of the last four models over *Blackboard* one is that the agent can directly exchange messages with another one.

13 Security

Security is a basic point for mobile agent messaging. Since mobile agent messages usually concern a number of different locations storing or processing them, there are lots of chances to steal, modify or read agent messages.

Blackboard, Broadcast and *Follower-Proxy* models are the less safe. Since *Blackboard* model is anonymous, any agent can read or write a message; it also gives the opportunity to a malicious agent to overwrite the messages left by another agent. Since *Broadcast* model sends a single message to all locations, many of them will have access to it. *Follower-Proxy* model has the same problem. Although it does not forward a message to all locations, it sends a message to all locations in the agent's route.

E-mail and *Home-Proxy* models are the safest ones, since message delivery occurs only between *home-place* and receiver agent. The security of mobile agent messaging can be improved, by using existing cryptography techniques. However, public and private key managing is the current goal and must be added to multi-agent system.

The table 1 highlights the main points discussed above that influence the decision on which model to choose for the exchange of messages in a mobile agent system. They include the number of agents (NA) in the application, the message type (MT) (direct or anonymous) required, possibility of failure (PF) in the model, number of messages needed to send a single message (MC) in an application, the amount of storage needed to support a single message (SC), complexity of design (CD) and implementation of the model, complexity of use of the model (CA) in the application and security level provided by it. On the basis of the considerations made before, each model has been assigned some points on a scale ranging from 1 to 3: 1 is the minimum, 2 is sufficiency, 3 the maximum. Total score column (TF) is the sum of each model's points, excluding the type of messages and the number of agents. This table points up that *Follower-Proxy* model is the less preferable, and that both *E-mail* and *Blackboard* are the best ones according to the messaging choice, direct or indirect. However, both *Home-Proxy* and *Broadcast* models are very close to the previous ones. Then which model should be used?

It is not easy to take such a decision. The considerations we have dealt with suggest that *Follower-Proxy* model should not be used to create a mobile agent system. Its message and storage costs are high, it has many possibilities of failing, it is difficult to be made safe, and complex to be designed and implemented. *Blackboard* model is interesting and should be used especially when messages must be sent between anonymous agents. Such approach excels in those points that were disadvantages for the previous model, but does not as far as the application

Table 1: Comparison among messaging models.

Model	NA	MT	MC	SC	CD	CA	PF	S	Total
Home-Proxy	Many	D	2	2	2	1	3	1	11
Follower-Proxy	Many	D	2	3	3	1	3	3	15
E-mail	Many	D	1	2	2	1	3	1	10
Blackboard	Many	A	1	1	1	3	1	3	10
Broadcast	Few	D	3	2	1	1	2	3	12

complexity is concerned. If anonymous communication cannot be used, or the application complexity is too wide to use anonymous messaging, *Home-Proxy, E-mail* and *Broadcast* models can be chosen, on the basis of the main application goals. If all the above-mentioned points are important, *E-mail* model should be chosen. If further caution is not needed for application security, *Broadcast* model is as valid as the *E-mail* one. It is clear that we cannot *a priori* decide which is the most suitable model, the choice having to be made according to the design instances peculiar of that system and on the basis of available resources.

References

[1] Andrews, G.R., *Concurrent Programming: Principles and Practice,* Addison-Wesley: Menlo Park, CA, USA, 1991.

[2] Andrews, G.R., Synchronizing resources. *ACM Transactions on Programming Languages and Systems*, **3(4)**, pp. 405–430, 1981.

[3] Hoare, C.A.R., Communicating sequential processes. *Communication of the ACM*, **21(8)**, pp. 666–677, 1978.

[4] Walker, E., Floyd, R. & Neves, P., Asynchronous remote operation execution in distributed system, *Proc. of the 10th International Conference on Distributed Computing System*, IEEE, 1990.

[5] Mishra, S. & Xie, P., Models for interagent communication and synchronization. *Proc. of the 2000 International Conference on Parallel and Distributed Processing Techniques and Applications*, Las Vegas: NV, USA, June 2000.

[6] Bernstein, A., Predicate transfer and timeout in message passing system. *Information Processing Letters*, **24(1)**, pp. 43–52, 1987.

[7] Duego, D., Mobile agent messaging models, *Proc. of 5th International Symposium on Autonomous Decentralized Systems*, pp. 278–286, 2001.

[8] Lange, D. & Mitsuru, O., *Programming and Deploying Java Mobile Agents with Aglets*, Addison-Wesley Longman Publishing Co. Inc.: Boston, MA, USA, 1998.

[9] Glass, G., *The ObjectSpace Voyager Universal ORB,* White Paper, pp. 1–12, www.objectspace.com, 1999.

[10] Lingnau, A. & Drobnik, O., *Agent-User Communication: Requests, Result, Interations*, in Lecture Notes in Computer Science (1477), Springer, pp. 209–221, 1998.

[11] Cardelli, L. & Gordon, D., *Mobile Ambients*, Foundations of Software Science and Computational Structures, LNCS No.1378, Springer, pp. 140–155, 1998.

[12] Murphy, A. & Picco, G.P., *Reliable Communication for Highly Mobile Agents, Proc. of the First International Symposium on Agent System and Applications and the Third International Symposium on Mobile Agents*, IEEE Computer Society, pp. 141–151, 1999.

Coordination

Davide Rizzo and Alessandro Genco
DINFO – Dipartimento di Ingegneria Informatica
Università degli Studi di Palermo

1 Introduction

Coordination is a basic aspect of mobile computing. It is in fact necessary to separate the treatment of single components from the management of their interactions [1]. The implementation of effective coordination tools implies several advantages, above all because one component's knowledge of others in order to communicate with them is limited.

Let us start by saying that two parts can form a complete *programming model*: a *computation model* and a *coordination model* [2]. The computation model enables a programmer to build a single computation unit: it therefore implies a single-threaded and step-at-a-time computation. Coordination model enables, on the contrary, to combine separate entities in a set of asynchronous mutually communicating ones. An entity is a program, a process, a thread or, generally speaking, any entity able to simulate Turing machine. Sets of entities are usually called *ensembles*.

Computation and coordination models can be either integrated into a single language, or remain distinct in two different languages, in which case programmers choose a specific reference model for each one. The second solution is probably the best one, since, if treated orthogonally to computation, *asynchronous ensembles* coordination problem can be solved more easily.

Coordination, at any rate, is not a simple exchange of information among active agents [3]. Each coordination language must allow the agents to communicate with other agents whose state is in continuous and unpredictable evolution [4]. That is why orthogonality is desirable. Nevertheless, alternative ways to orthogonality – hence to coordination and computation – have often been considered. Most traditional distributed systems, for instance, are based on RPCs (remote procedure calls), that are however in contrast with parallel applications.

Each coordination model is formed by three elements [5, 6]:

- *coordinables*: the entities whose interactions are regulated by the model (e.g. Unix processes, threads, etc.);
- *coordination media*: the abstractions allowing interactions (e.g. traffic lights, monitors, tuple spaces, etc.);

- *coordination laws*: they can be defined in terms of *communication language* (language for the exchange of information and data structures) and *coordination language* (set of interaction primitives).

After having defined such elements, it should be added that coordination could be achieved by mechanisms of identification of the nearest entities, of exchange of information, of synchronization, etc. Mechanisms can be:

- *explicit*: an entity refers to another one when it wants to send it a message;
- *implicit*: coordination is completely transparent to entities.

An important aspect of coordination is the ability to determine *who else is around*. A minimum knowledge of the other participants to computation is in fact required to mobile units in the majority of models, though they are often defined as *transparent*. This knowledge, on the other hand, could be used to optimize performances.

2 Coordination in mobile agent systems

Distributed applications are usually developed around a set of processes statistically assigned to specific execution environments, mutually cooperating, without any network's direct control. Mobile agents paradigm, on the contrary, defines applications as a set of active entities able to change their execution environment [7], by moving to other nodes during their activities.

In mobile agent applications, one of the basic activities is *coordination* among agents and the entities they meet during their execution, be they other agents or simple *resources* (hardware and/or software ones) available on the execution environment. Anyway, both mobility and the scenario extent introduce different problems than traditional ones based on RPC paradigm. The scenario becomes more complicated if we consider applications where the number of agents involved is very high. On the one hand, in fact, specific languages such as FIPA and KQML succeed in a way or another in solving coordination problem through communication, while, on the other hand, as far as interoperability problems are concerned, solutions such as the one proposed by CORBA claim a wide level of maturity. Such solutions, however, concentrate above all on peer-to-peer communication and lack a global vision of the agents/execution environments ensemble, hence the need for a specific coordination model for mobile agent systems. A coordination model is necessary:

- to avoid anarchy and chaos in order to achieve a common goal;
- to exchange resources and distributed information;
- to manage dependences among the agents' actions;
- to increase overall efficiency.

3 Coordination models

An application can be composed by several mobile agents cooperating to complete a task, for which they need to coordinate their activities. It is important to establish:

- how the agents communicate and synchronize their activities;
- how the agents interact with execution environments.

The first point is about *inter-agent coordination*. An application can be formed by several agents (if necessary mobile ones) cooperating in order to pursue a goal, for which they must absolutely synchronize their work, as well as, generally speaking, exchange data and knowledge.

The second point is particularly critical when mobile agents must move through the Internet to access to remote resources and services located in network nodes. As a matter of fact, when agents migrate to a node, they need to access to the resources and services available in the new environment hosting them. We refer to the latter as *agent-to-hosting-environment coordination*.

3.1 Taxonomy of coordination models

The basic differences between the various proposals in available literature turn around the concepts of *spatial* and *temporal coupling* [8]. In particular:

- *spatially coupled* coordination models need that involved entities share a common name space, while *spatially uncoupled models* represent anonymous interactions;
- *temporally coupled* coordination models lead to a form of entity synchronization, while *temporally uncoupled* ones lead to asynchronous interactions.

Four large model families ensue [9, 10, 11], hereafter called *direct*, *blackboard-based*, *meeting-oriented* and *Linda-like* (fig. 1).

Spatial (Name) \ Temporal	Coupled	Uncoupled
Coupled	Direct *Odissey, Agent-TCL*	Blackboard-based *Ambit, ffMain*
Uncoupled	Meeting-oriented *Ara, Mole*	Linda-like *Jada, MARS, TuCSoN*

Figure 1: Taxonomy of coordination models.

We bring as an example a simple Internet data retrieval application [10], where an agent is supposed to be sent to a remote site to analyse html pages and returns the URLs containing a specific keyword. The agent clones itself for every remote link found on the analysed pages and sends there clones. *Inter-agent* coordination is necessary to avoid multiple visits of the same agents to the same site, while *agent-to-hosting-environment* coordination establishes a specific protocol to have access to the site's information.

For the sake of completeness, before going on with the detailed analysis of the models just dealt with, we must mention a second taxonomy of coordination architectures, proposed by Lee [12] and accepted in some fields of research: according to such scheme, a coordination model can be classified, on the basis of the implementation techniques used, into one of the following four categories:

- *organization structuring*: organization patterns are *a priori* defined to implicitly establish responsibilities, abilities and connectivity of all agents involved. Middle/long-term relations are fixed among entities that will have to coordinate their activities to achieve a certain task – master/slave and client/server paradigms are typically used;
- *contracting*: based on the so-called Contract Net Protocol, on the basis of the metaphor of a decentralized market structure;
- *multi-agent planning*: agents define a decisional plan describing all actions and interactions needed to achieve pre-arranged goals. Planning can be centralized or decentralized;
- *negotiation*: thought of as "... a group of agents' communication process in order to reach a mutual agreement of some kind" [13]. Game theory-based, plan-based and human-inspired negotiation models are classified in this field.

3.1.1 Direct coordination

In *direct* coordination models (fig. 2), agents begin to communicate by explicitly naming the partners involved (spatial coupling). That intrinsically implies a form of temporal synchronization (temporal coupling). Two agents must agree on a common communication protocol, typically a peer-to-peer one. Coordination with resources available in the host (agent-to-hosting-environment coordination), on the contrary, makes use of a client–server paradigm.

Although it has interesting and convenient features in particular conditions, overall adoption of such coordination model presents some difficulties, first of all because the frequent interactions required need highly stable network connections, making thus communication highly dependent on network congestion factor. Furthermore, communication among entities residing in wide networks (such as Internet) requires particularly complex routing protocols that could affect tasks execution in terms of performance (the agent would strongly depend on network latency).

Finally, since mobile agents applications are intrinsically dynamic (through dynamic creation of agents), it could be difficult to adopt a *spatially coupled*

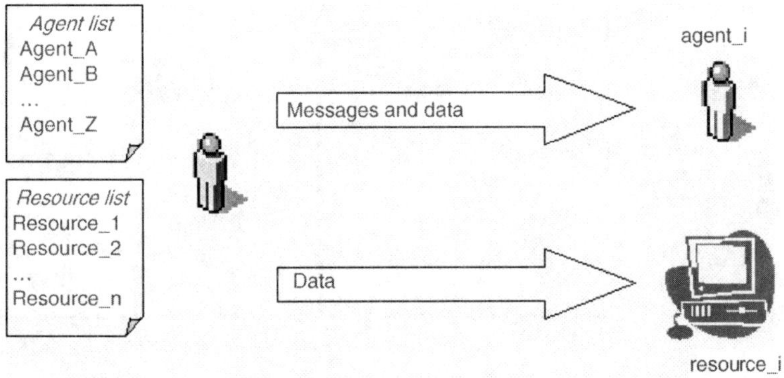

Figure 2: Direct coordination: an agent communicates only with agents and resources known thanks to constantly updated lists.

model in which the partners involved in communication should precisely identify each other. Besides, agents meant as autonomous entities cannot *a priori* know how many components are included in their application (in the case of multi-agent systems) and, however, even if they are known, the fact they must coordinate through synchronization goes against the autonomy of agents that would thus depend on others.

Anyway, in the presence of wide scenarios, direct coordination can effectively prove to be winning if it is applied to access to local resources (agent-to-hosting-environment coordination): a local server represents the manager of such resources and agents interact with it through client–server paradigm.

The best known among mobile agents systems using direct coordination model are *IBM Aglets* and *Agent Tcl*.

3.1.2 Meeting-oriented coordination

In *meeting-oriented* model, agents can interact without the need of explicitly naming their partners: interactions occur in a specific context acting as a *rendezvous* for all the agents involved. Apart from *rendezvous* abstracting the server's role in an execution environment, an active entity must necessarily assume the figure of supervisor (or *initiator*) to actually open the meeting space (fig. 3(a)). A meeting often takes place in a particular execution environment, in order to avoid problems linked to remote communications, and only local agents can participate. Clearly, since the agents must know both the names used in the meeting and the events forcing them to join the meeting itself, a complete spatial uncoupling cannot be carried out (fig. 3(b)).

Meeting-oriented model partly solves the problem of the exact identification of partners, but introduces a strict form of synchronization among agents. If we think that agents' scheduling and position can hardly be anticipated in several applications, the risk of losing interactions is very high.

Figure 3: Meeting-oriented coordination: a group of agents having to interact meet in a meeting point (a) and, after a login phase managed by a supervisor entity, obtain explicit pathnames of the participants with which to interact (b).

In test applications, in order to avoid several agents visiting the same site, an extra agent can be added for each site visited: when an agent has explored a site, it creates an agent acting as meeting server, forcing him to stay in that site waiting for other possible agents arriving later. Such a solution is not however easy to achieve, mainly because the agent acting as meeting server is supposed to have the possibility of remaining active on the hosting site.

Meeting-oriented coordination is implemented in *Ara*: an agent takes on the role of meeting server setting up a meeting point in a hosting environment; incoming agents can participate and coordinate each other.

3.1.3 Blackboard-based coordination

In *blackboard-based* model, interactions occur through shared *data spaces*, precisely called *blackboard*, used as common stores for the storage and retrieval of messages. If agents keep to a common message identifier and communicate through blackboard, they are spatially coupled.

The basic advantage of this model [14] is due to complete asynchrony among agents: messages are left on blackboards without knowing if and when recipients read them (fig. 4). We can easily guess the advantages of their use in wide scenarios where agents' position and scheduling cannot be monitored nor guaranteed. Moreover, every agent's interaction being forcibly realized through local blackboard, hosting environments can easily control what happens, increasing the degree of safety that is one of the weak points in traditional coordination models.

There are several available solutions that fully exploit blackboard-based coordination model: in *Ambit*, a recent model for mobile computing, agents can read and write messages on the local blackboards of the sites hosting them; *ffMAIN* agents, instead, interact – both with other agents and with the

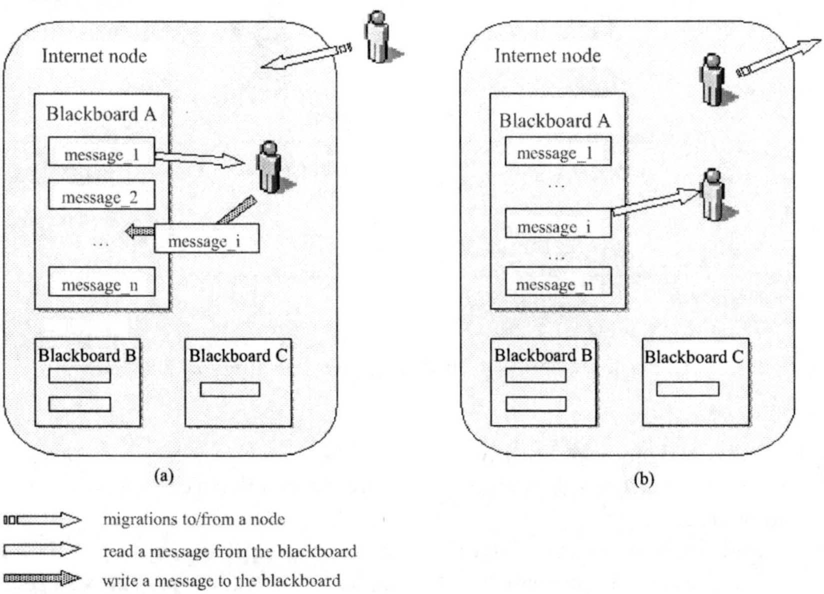

Figure 4: Blackboard-based coordination: an agent reads and writes messages on one of the blackboards available in the node (a); a second agent can asynchronously read the message left before (b).

environment hosting them – through a data space accessible through an HTTP protocol, thus providing the model with a weak form of spatial uncoupling.

3.1.4 Linda-like coordination

Linda-like coordination model has the greatest success at research level [11], although mature architectures implementing it are not many. Linda-like coordination is very close to blackboard-based one, though an *associative* kind of mechanism is added: associative blackboards, commonly called *tuple space*, reach a form of full uncoupling, since they do not need temporal synchronization nor mutual knowledge of the entities that must be coordinated (fig. 5).

The concept of associative blackboards has been implemented by *Jada* system as well as by *JavaSpace* architecture: the so-called *ObjectSpace* is the abstraction used by mobile agents to save and retrieve associatively references to objects. Furthermore, agents can create ObjectSpaces tested to interact without the hosting environment interference.

Linda-like model can be extended by adding the so-called *reactivity* [15]: a reactive tuple space is not only anymore a simple and without history store with a neutral associative mechanism; providing it with programmable reactions influencing accesses, it assumes the connotations of a space with its own peculiar

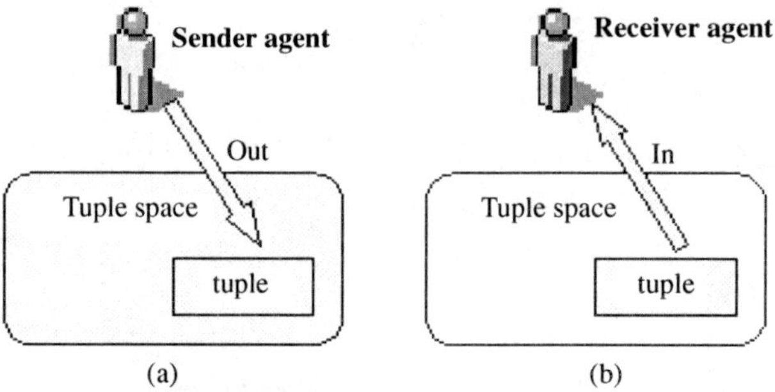

Figure 5: Linda-like coordination through tuple writing (a) and reading (b).

state, better and more specifically reacting to the agents' requests. Reactions can, moreover, access tuple space, modify its content and influence the agents' semantics to interact with it.

The advantages deriving from the adoption of reactivity are many [7]: reactions can be used to implement specific local policy of interaction between agents and execution environment, in order to reach a better control and protect the environment's integrity itself from potentially malicious agents. Furthermore, reactions can dynamically adapt interaction semantics to the specific features of the execution environment, thus simplifying the agent's task programming. Still more, a *reactive space* enables to state precisely the rules of inter-agent communication as if they were high-level instructions, thus clearly separating algorithmic phase from the purely coordination one in the agent's programming. On the other hand, the application possibility to define its own tuple space contributes to separate clearly the concepts concerning algorithms from those merely linked to coordination:

- agents are charged to achieve their task according to a certain algorithm,
- reactions represent the particular coordination rules for that specific application.

Thus, the agent does not need to know the peculiar characteristics and the services available in the hosting environment; it simply accesses to local tuple space using the same coordination style both for inter-agent one and for the environment's reactive behaviour. Furthermore, reactive tuple space can considerably help and increase control, as well as protect environment's integrity from potentially dangerous agents.

If, on the one hand, several proposals in the field of coordination in general aim at adding reactivity to the flat Linda-like model, so far the proposals directly linked to the world of mobile agents are few: *TuCSoN* model defines programmable *tuple centres* for the coordination of *knowledge-oriented* mobile agents; tuple space defines a Linda-like interface and reactions are programmed as first rate logic tuples. Another instance easily applicable to mobile agents technology is the *PageSpace* project, which defines a Linda-like model enriched for distributed Web applications.

3.2 Context-dependent coordination

In Internet applications based on mobile agents, it is convenient to abstract the network as a variety of execution environments (e.g. nodes, domains, etc.), and develop applications in terms of agents fully aware of the distributed nature of goals and able to move from one environment to another to have access to the allocated remote resources. Agents, during their execution, need to interact, communicate and synchronize their activities both with other agents (inter-agent coordination) and with the resources available in the hosting environment (agent-to-hosting-environment coordination). As we have already seen, both coordination modes can be easily carried out through the use of middle coordination such as blackboard, tuple space, etc., each of them associated to a particular environment. Such infrastructures fit well to the mobile nature of agents strengthening the principle of locality. It is also true, however, that when agents enter wide contexts, moreover heterogeneous and not easily predictable ones, mere inter-agent interactions display some limitations, among which the fact that the agents migrate to different execution environments during their execution, coming into contact with different kinds of data, arranged in different spaces and managed by different agents according to each case. This form of context-dependence is intrinsic to agents' mobility, independently from interaction space. That is why the need is felt for more sophisticated *context-dependent* coordination forms [16] based on the active role of context in a way or another:

- every execution environment has its own characteristics and safety policy, and could need in some way to impose specific laws to the agents it hosts, thus needing an environment-dependent form of coordination;
- the agents of a particular application could need to follow particular methodologies to achieve the goal they have been programmed for independently from the characteristics of the environment hosting them, thus showing the need for an application-dependent form of coordination.

If interaction spaces are enabled to dynamically program their reactions to the agents' events, such spaces can behave in a different way according to the agent involved in that specific interaction. This feature thus can be used as a context-dependent form of coordination, and in particular:

- the site's administrator can adapt interaction space behaviour to strengthen – transparently to the agent – specific safety policy;
- agents can dynamically adapt their behaviour in a site according to its laws.

Before analysing these concepts in detail, let us take into account an interaction architecture both local and uncoupled: interactions are limited to a specific space associated to the execution environment and, besides, the use of an interaction space enables communication among agents without the need to know their precise name and position. A local and uncoupled model, furthermore, suggests that application can be thought of in terms of mobility.

When an agent migrates to a new execution environment, its interaction space, and therefore its perception of the surrounding world, changes according to its movement: the result of a given interaction in the new site will probably be different from the one it was used to in the previous site. This context-dependent form of coordination is intrinsic to agent's mobility and to the adoption of local and independent interaction spaces. However, if space is reduced to a simple infrastructure to store and retrieve data, messages or tuples, the simple context-dependent form just described becomes for the agent quite complex to be dealt with. First of all, to coordinate its own actions with the other agents residing in an unknown environment would require the same kind of solution as to the typical problems met in the so-called *open systems*, i.e. unpredictable behaviour, a certain variety of languages and protocols, heterogeneous interfaces and so on. Secondly, each environment has its own specific characteristics and it can sometimes force the hosted agent to coordinate its activities in a specific way for safety reasons and for the control of allocated resources.

In the case of static and not programmable interaction spaces, the problems just mentioned would lead to the creation of very heavy and complex agents, requiring costly maintenance to adapt to the frequent changes of the environments they would be destined to. Instead we could deal directly with interaction space, in order to manage it in accordance with its distinctive features. Thus, such space drops its static quality to become completely active and programmable.

For example using a model with programmable tuple spaces, we could:

- characterize the various kinds of accesses according to the agent's identity;
- express a new behaviour (reaction) the space should assume as a response to such accesses.

It will be noticed that such model is nothing but the one previously shown as reactive Linda-like, even if from the point of view of context-dependent coordination and not simply as coordination media. Therefore, the adoption of a programmable interaction space leads to the scenario described in fig. 6.

Running agents in a certain site, whatever is their application, are subject to local environment laws. Furthermore, the agents of a certain application can make use of their own coordination laws on the site they visit in order to influence all other agents' coordination activities in the same application.

3.3 Environment-dependent coordination

Agents, although they have access to reactive space, be it a tuple space or not, always with the same interface, can nevertheless show variations in the semantics of their coordination activities (as well as in the perception of the surrounding environment) due to the specific behaviour programmed in local tuple space. In other words, the same interactions can have different effects according to the site in which they are executed. This feature can be used both to incorporate safety policies and to help agents that enter a new environment without having

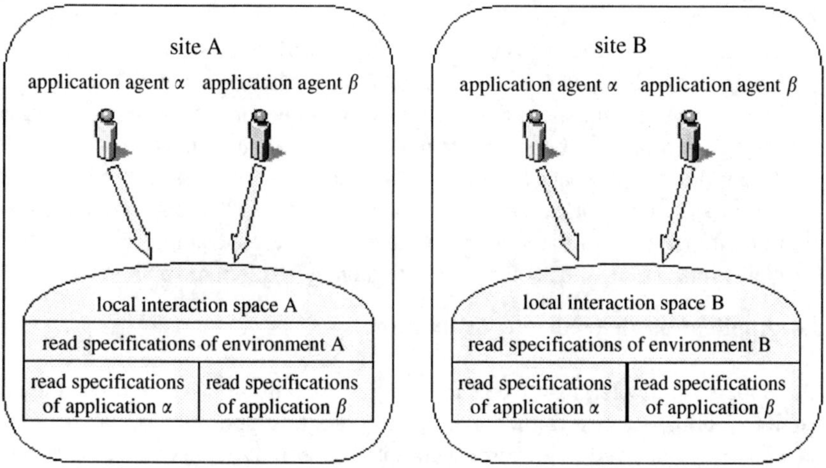

Figure 6: Context-dependent coordination.

preliminarily and explicitly agreed on all the characteristics of the means at their disposal.

Reinforcing safety: when a site is available for the execution of mobile agents, it must be aware of the fact that possible malicious agents could try to undermine its data and resources integrity that must therefore be protected against unauthorized accesses. Agents, on the other hand, can be uninformed of such safety policies and therefore their interactions will probably set off a high number of security exceptions that they must deal with by themselves. Without entering into the merits of safety in distributed systems, particularly mobile agent ones, we can however say that if all interactions in the environment are mediated by a programmable tuple space, the problems just mentioned can find a powerful and elegant solution: the administrator, in fact, can plan tuple space's behaviour so as to make the access not dangerous for the integrity without delegating to the agent the solution of the exceptions it raises. In the usual example application, if an agent in search for html files tries to select the tuples representing them, while it has only the right to read them, a specific planned reaction in the tuple space could deliver the desired tuple to the agent without taking it away from its space, and without any exception.

Managing heterogeneity: as we have already said, agents must necessarily tackle a heterogeneous world. In the case of a reactive tuple space, we must nevertheless look at the problem from a completely different perspective: being a tuple space programmable to react to agents' accesses, it can make such requests homogeneous, at least for the agent's view and within the limits of local execution environment. Besides, agents can perceive tuple space as if it were in line with its expectations, without having to double space representations or force them to a better manageable representation.

Support to open interactions: the use of programmability of interaction space to adapt to heterogeneous exigencies naturally extends to those cases in which a site is supposed to be open to the execution of agents belonging to different associations and applications, possibly heterogeneous ones in terms of supported languages and protocols, but once again with the need to coordinate with each other. Once more, a tuple space (or any programmable interaction space) can be useful if used as a mediator, so as to support coordination activities among agents of different nature. Still more, a reliable site could be used as impartial referee for interactions among too selfish agents belonging to different applications.

3.4 Application-dependent coordination

Developers can exploit reactive space programmability in various ways: they can use it to facilitate access to information on a site, to support the exchange of large data among agents and to implement complex coordination protocols. More generally, they can use reactive space programmability so as to adapt the model of general coordination to their specifications and needs. Obviously, since such changes to the behaviour of the hosting environment are made to meet specific application features, a trespassing mechanism must be provided to be sure that an agent's behaviour influences only the same application agents, not indiscriminately all the ones sharing the same interaction space but belonging to different applications.

Including coordination protocols inside interaction space rather than into the agent's logic helps to streamline the planning/designing phase/stage as well as the application and its agents' maintenance, since the administrator of the execution environment will have the task to dictate the behaviour rules replying to agents' accesses, as well as to update and modify them according to precise exigencies – security ones or more simply control ones.

Looking again at the development of a mobile agent application from the point of view of context-dependent coordination, rather than from the traditional one of a static model (the direct, blackboard-based, meeting-oriented or Linda-like one), it can be easily realized how the availability of an infrastructure based on the programmability of local interaction spaces can have a positive impact on agent-based software engineering. Such model, in fact, naturally induces to develop the application by neatly differentiating mere *intra-agent* aspects – linked to specific computation roles of the agent – from the *inter-agent* ones – linked to the interactions with other agents and with the execution environment. The development of an application can actually proceed simultaneously on two main lines:

- intra-agent engineering:
  - what kind of information must my agents retrieve in the site they visit?
  - how must they analyse information and draw useful data?
  - how are such data returned to the user?
- inter-agent engineering:
  - how do agent influence each other as they visit the same site?
  - what kind of information can they exchange through programmable interaction space?

To sum up, through the use of a context-dependent coordination model emerges a sharp distinction of roles between sites' developers and administrators: when a new type of application is being introduced in the Internet, site's administrators can develop and implement all the environment-dependent coordination laws they think necessary to facilitate the execution of the new kind of agents and at the same time protect themselves against improper practices.

The separation of concepts during the planning stage is kept in the final code too: the code that implements coordination laws (the environment-dependent one or the application-dependent one) is as a matter of fact separated from the agent's one, and it furthermore can be added, modified or extended in a totally modular and independent way.

So far, however, it lacks a strong interaction of such coordination infrastructures with the agent communication languages (ACLs) enabling the agents to mutually interact through high-level expressions.

4 Coordination languages and Berlinda

As already seen, the use of a proper coordination model, to be chosen on the basis of the application's specific exigencies, is essential in the development of multi-agent systems. The alternative choice to use a simple coordination language, rather than a complete architecture, can anyhow prove to be enough if not even better in those cases in which the number of entities involved in the interactions is small or easily manageable. An agent coordination language defines the kind of elements, the operations to create and destroy elements, as well as the activities typical of coordination that are needed to manipulate coordination media. In this field, therefore, agent coordination is expressed through high-level linguistic means, and the ensuing behaviour is determined by the semantics of the particular language chosen and implemented in the underlying architecture. Nowadays there are many available solutions making use of a coordination language: in the *Object Management Architecture*, agents are manipulated through CORBA-API; in approaches to parallel programming such as PVM message exchange is used through a specific API; KQML makes use of typified messages defining at the same time their semantics. Rather than presenting an overview of solutions and typologies of languages definable as coordination languages, which is not part of the aims of this document, Berlinda architecture [17] is briefly outlined. It is a model defining a set of *foundation classes* common to the various coordination languages. The definition of the so-called *meta-coordination* language, necessary to facilitate communication through multiple and heterogeneous environments, is also given here. To this purpose, CORBA can be thought of both as coordination language and as meta-coordination one.

To sum up, Berlinda approach consists in the definition of common formats and specific APIs for agents integration, associated to the use of a meta-coordination language to deal with interoperability and interface with multiple systems: it is defined as a highly abstract coordination model in the broad sense, later instanced in a specific coordination language among the supported ones. The common structure of the various coordination languages that can be expressed and implemented through Berlinda is not limited to the multisets used as coordination media: *shared media* are also used, central abstractions for location concept (a multiset, an interface store or a message virtual space are all examples of shared media). Berlinda's basic concepts, completely corresponding to the ones already introduced in some cases, have different names in coordination architectures:

- *the coordination medium*, the structure shared by agents, provides manipulation operations in the form of coordination language. In particular, it is abstracted from the concrete data structure, providing though *multiset medium* implementation;
- the medium is a collection of *elements*, vectors whose fields belong to the type set supported by the system;
- elements can have a *signature* providing meta-information on the current element;
- elements provide a *matching function* relating them according to the coordination language semantics; it can be some form of Linda-like pattern matching mechanism or degenerate in the simple identification of the involved entity;
- *agents* are execution threads using the manipulation operations given on coordination medium.

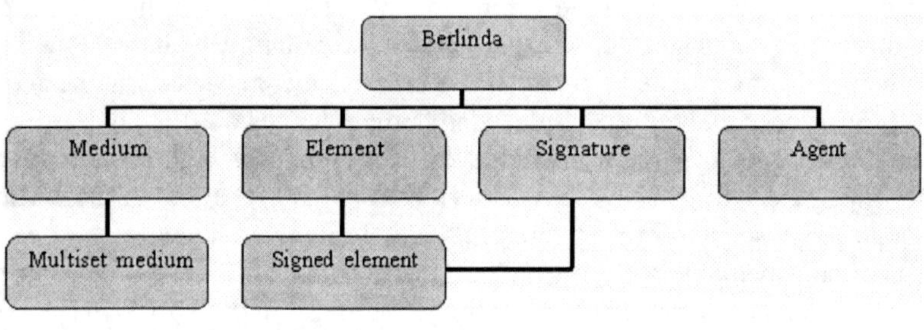

Figure 7: Berlinda model.

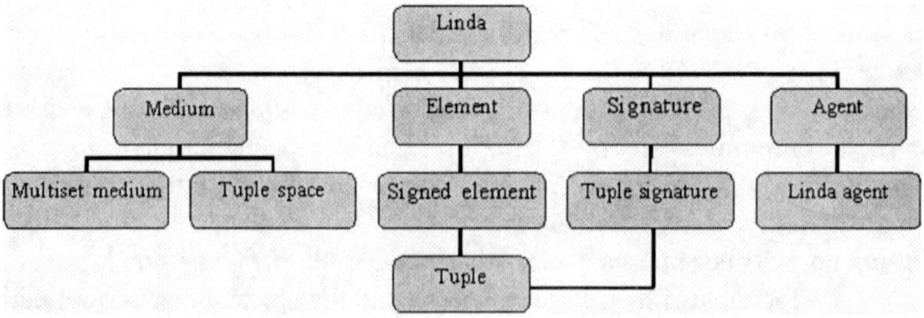

Figure 8: Linda language within Berlinda model.

These five points cover all the aspect already studied, with a different approach, in coordination models. The following schemes (figs. 7 and 8) show Berlinda's general architecture and Linda system, seen as a coordination language within Berlinda.

5 Implementation of coordination models

We will deal now with some of the most widespread solutions in the field of mobile agent architectures, focusing exclusively on coordination peculiarities rather than on the whole framework.

5.1 IBM Aglets

One of the most widespread mobile agent systems is the one known as Java Aglets and developed by IBM Tokyo Research Laboratory [18]. It is an extension of Java language with support to *weak mobility*. Agents, called Aglets, are nothing but Java threads.

Aglets communicate by exchanging messages in the form of *Message* class objects, although the system is projected not to allow them to have direct access to another agent invoking its methods, even if they are public; on the contrary, every Aglet is associated a *proxy*, i.e. a sort of container hiding its inside (the Aglet, precisely) ensuring though that the operations are carried out through a proper interface (fig. 9).

A message is sent through a call to the proper method on the Aglet's proxy; that is why the sender must necessarily keep trace of the movements of the receiver Aglet.

The presence of proxy is also useful to reduce, even if a little, spatial and temporal coupling of coordination model, since the former is actually divided into two parts, one concerning the sender Aglet and the other concerning the proxy of receiver Aglet; similarly the latter is made lighter because it is limited only to the synchronization of the proxies involved in the message exchange.

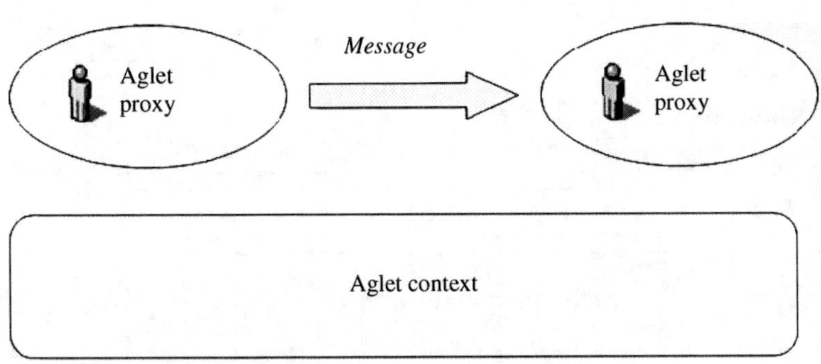

Figure 9: Message-passing coordination IBM Aglets system.

The Aglets system furthermore supports various types of native message to invoke specific proxy methods:

- *sendMessage*, to send a message and stop current execution until the addressee has completed reception.
- *sendAsyncMessage*, to send a message without stopping current execution. The result will be given later.
- *sendOnewayMessage*, to send a message without stopping current execution; it is different from the previous one because it does not give back any value.

Aglets provide the notion of *context*, which abstracts the execution environment and has a set of basic services, such as the creation of an Aglet. An interesting service is the acquisition of the list of Aglets present in a certain context. Generally speaking, an Aglet's proxy can be recovered by specifying a single identifier for each of them, the *AgentId*: such operation can be achieved in different ways, such as by asking the context, that operates as a sort of local name server, through the *getAgletProxies*() method; otherwise the *getProperty*(*String*) method could be used, ensuring that the system obtains a reference through the *setProperty*(*String,Object*) method.

Moreover, each context allows the agents to have access to their local resources by activating agent-to-hosting-environment coordination.

Finally, being the system merely written in Java, it can make use of language RMI (remote method invocation) mechanisms, acting once again in the form of spatial and temporal coupling.

5.2 Ara

Developed at the University of Kaiserslauten, the Ara (agents for remote action) system uses the same basic solution concerning portability and safety: agents are executed on a Virtual Machine, typically an interpreter and a run-time system

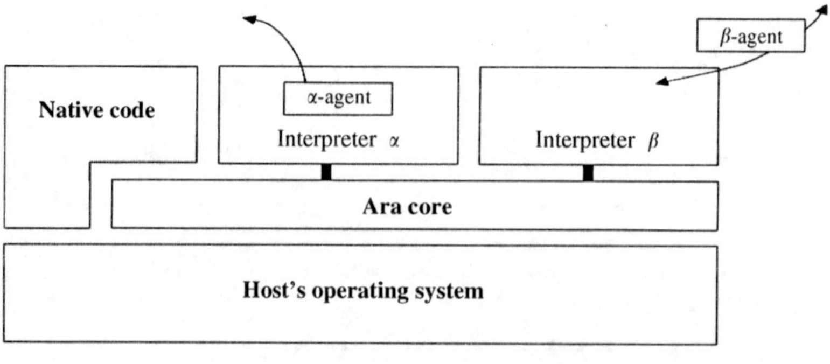

Figure 10: Ara system's architecture.

(fig. 10), both needed to hide the details of the host system architecture and to confine agents' actions to a closed environment. In practical terms, Ara agents are programmed with some kind of interpreted language [19] and executed through a specific interpreter, supported by the so-called Ara core. The relation between core and interpreter has the particular aim to isolate essential characteristics (such as to catch the C's specifics of an agent's state from the ensemble of language C) in the interpreter and concentrate in the core all functionalities independent from the adopted language. In order to support compatibility with existing models, Ara does not prescribe a programming language for its agents, but an interface to which the desired languages can be coupled: this feature enables the use of different interpreters associated to a single core, which makes the latter's functions evenly at the disposal of all the solution chosen. Apart from migration support, the main functions of core include management, communication, persistency and safety mechanisms.

As to coordination choices, Ara emphasizes local interactions among agents through two simple mechanisms: synchronous message passing and tuple space. In the first case, the core introduces the concept of *service point*, symbolically called rendezvous, where agents can interact both as client and as server through the exchange of messages with arbitrary format in the form of request/reply. Every request is marked with the client agent's name that later is used by the server agent to reply.

The other communication mechanism among agents is tuple space implemented by a system thread, accessible as both server and service point. Agents can leave there structured data to be recovered/recalled asynchronously by addressees. Anyway, independence from language typical of Ara system compels to the use of an interface in order to make use of such interaction space, typically a sophisticated mapping language as CORBA, which makes the general mechanism more complicated than the simple service point solution.

5.3 ffMAIN

ffMAIN (Frankfurt mobile agents infrastructure) project is part of a wider research context in mobile communication; it has been developed with the aim of creating a platform enabling maximum flexibility both in supported applications and in agents implementation methodologies. At the beginning based on http, base application has later moved towards TclHttpd system and Perl technology, and implementation efforts have focused on a combination that had to be independent from adopted language (Java, Tcl, etc.), but linked to standard http protocol.

At the basis of ffMAIN architecture [20] is the notion of *server*, i.e. a program (as a mail server, an FTP server, etc.) on every computer accessible to agents and entrusted of their local execution. Basic tasks take into account the agent's acceptance by other servers and users, the creation of suitable execution environments, the supervision of agents, and their transfer and communication arrangements. Therefore, a specific execution environment is created for each running agent on a server, and local resources are used through a suitable interface between the agents and the server itself, leaving the agents the client role. As far as interactions are concerned, agents communicate with each other through what is called *information space,* i.e. a blackboard local to each agent server containing triples (three-elements non-associative tuples) formed by an *item's key*, an *access control list* and a *value,* stored in complex or simply binary form, and not interpreted by the agent server. Agents can both destructively and non-destructively read or write in the information space; operations are atomic and serialized to ensure the lack of *race condition* and, in general, data inconsistency. For each element, operations can be carried out by a specific agent, a group of agents, or by all the agents that want to, according to access control lists.

The use of information space obviously leads to temporal uncoupling among agents and very simply enables local implementation of multicast schemes. Such spaces can finally be reached from the outside, for example through a WWW browser. It is enough to associate objects with a MIME type that, once interpreted, is given back to the browser, thus increasing more and more this technology's possibilities of use within the Internet.

5.4 JavaSpace

JavaSpace technology [21], developed by Sun Microsystems, is a simple unified mechanism for communication and dynamic coordination as well as for the exchange of Java objects. JavaSpace makes use of RMI and serialization in order to have the so-called *distributed persistence* and enable the achievement of distributed algorithms. JavaSpace is therefore a new platform for distributed systems remarkably simplifying their design and achievement. A second goal achieved by this technology is to make the application client-side completely achievable in Java, with a moreover limited number of classes. JavaSpace technology is strongly influenced by Linda. *JavaSpace services* are based on the concept of *entry.* An entry is *typed* group of objects. All entries are instances of Java classes implementing a certain Java interface (net.jini.core.entry.Entry). The concept of entry perfectly corresponds to that of *tupla* typical of Linda. The entries fields can be set to *values* (Linda's *actuals*) or stay *wildcards* (Linda's *formals*). *Write* (Linda's *out*), *read* and *take* (Linda's *rd* and *in*) operations are allowed. JavaSpace can also be asked to *notify* when an entry that *matches* a certain template is written. That is done with an events model contained in net.jini.core.event.

Let us try hereafter to make the concept of *match* clear. Two entries *t* and *u* match if:

- *u* is an instance of *t* class or of one of its super-classes; that extends the Linda model because it also allows matching among different tuplas belonging to the same hierarchy;
- *u* fields representing primitive types (e.g. char, integer) coincide with *t* ones;

- *u* fields not representing primitive types (e.g. objects) are the same, in their *serialized form*, as the *t* ones; two Java objects have the same *serialized form* only if their *instance variables* have the same values;
- a null value in *t* matches a null value in *u*.

Apart from the semantic ones we have just focused on, JavaSpace has important differences from Linda: contrariwise to Linda systems, JavaSpace uses *rich typing*. It means that entries themselves, not only their fields, are typed. In other words, two entries having the same field types but different data types are different entries. For instance, an ensemble of two numeric fields can be both a point and a vector. While in Linda two tuples representing a vector and a point would be formally the same and therefore completely undistinguishable, in JavaSpace the entries corresponding to the point and the vector would belong to two different classes. Being objects, JavaSpace entries are associated with an ensemble of methods, namely a certain *behaviour*; entries fields are also Java objects. Consequently, all the systems created through the use of JavaSpace are completely object-oriented.

In JavaSpace, operations are invoked on a Java object implementing *JavaSpace* interface. Such interface is not remote: every implementation of a JavaSpace service exports objects implementing locally on the client the *JavaSpace* interface and communicating with the real JavaSpace service thanks to an implementation-specific interface. Any *JavaSpace* method implementation can communicate with a remote JavaSpace service. Let us examine now the essential aspects of operations:

- *write*: a write corresponds to Linda's *out* and puts an entry in a JavaSpace service, i.e. in a space. *Write* is invoked by passing an *Entry* object as parameter. This operation gives back a *Lease* object corresponding to a *lease* expressed in milliseconds. When the lease expires, the entry is removed from the space;
- *read*: two kinds of reading can be made, i.e. there exist two ways of looking for an entry matching a certain Entry template. In both cases, the operation gives back a copy of the found entry, or null if it does not exist any entry matching the template. *ReadIfExists immediately gives back* an entry or null, and waits a certain timeout only if there are conflicts with the tuples possibly matching the template. Ordinary *read*, however, waits until a matching entry has not been found, unless the timeout expires;
- *take*: this operation, corresponding to Linda's *in*, works exactly like the *read* (*take* vs. *takeIfExists*) one, with the difference that entries are removed from the JavaSpace service.

Operations on a space do not consider in advance any pre-arranged treatment order. An inter-thread order can be achieved only through an *ad hoc* coordination mechanism. For example if two threads T and U invoke a *write* and a *read* in turn on two matching entries, the *read* could not find any entry even if the *write* returns before it. Only if T and U cooperate to ensure that the *write* comes back before the *read* starts, U will have the possibility to read the tupla entered by T (unless there are *takes* executed by a third thread).

JavaSpace has the following two innovative aspects:

- *Snapshot*: every time the same entry is used to execute an operation, exactly the same serialization process is repeated. *Snapshot* method is a means to reduce the impact of a certain entry repeated use.
- *Notify*: a *notify* request invoked on a template "records" the interest for the entries matching the template itself, among those that later will be entered in the JavaSpace service. When every entry of this type arrives, notify will take place, in the form of exception.

An important concept in JavaSpace is that of *transaction*. JavaSpace API makes use of a net.jini.core.transaction package to support the achievement of atomic transactions grouping together multiple operations in a bundle acting as an atomic operation. *Read, take* and *write* operations having the transaction parameter set on null act as if they were part of a transaction containing just the operation itself.

5.5 Mars

Developed at the University of Modena in the context of MOON [22] research project, Mars (mobile agent reactive spaces) system implements a Linda-like coordination architecture for Java-based mobile agents in execution on heterogeneous networks (Internet) [23, 24]. It is assumed that each node in the network hosts a server mobile agent able to accept and execute arriving agents as well as to store temporarily a reference to local tuple space: when an agent arrives, the first operation will always be the request of such reference to local server (fig. 11).

Mars is composed of various independent tuple spaces [5], available for both inter-agent coordination and agent-to-hosting-environment one. Every tuple space is associated to a node and can be accessed by locally executed agents. In order to integrate Mars with any mobile agent system the only requirement is that the server – accepting and executing the agents entering a node – gives the agents a reference to local Mars tuple space.

Mars follows JavaSpace specifications, by now almost a standard as far as tuple management Java interfaces are concerned; actually, Mars tuples are Java objects whose instance variables represent the fields of the tuple itself. The Linda primitives provided concerning the basic operations executable on a tuple space are:

- *write*, which writes a tuple, given as first parameter, in the tuple space;
- *read*, which reads a tuple matching the given template;
- *take*, which works as read except for the fact that it physically takes a matching tuple from the tuple space;
- *readAll*;
- *takeAll*.

Tuples used as templates through *read, readAll, take* and *takeAll* operations can have fields with both explicitly defined values and null ones. A tuple T will match (matching tuple) a template U if the defined values of U are the same as or matching the values of T, following the rules of JavaSpace pattern matching.

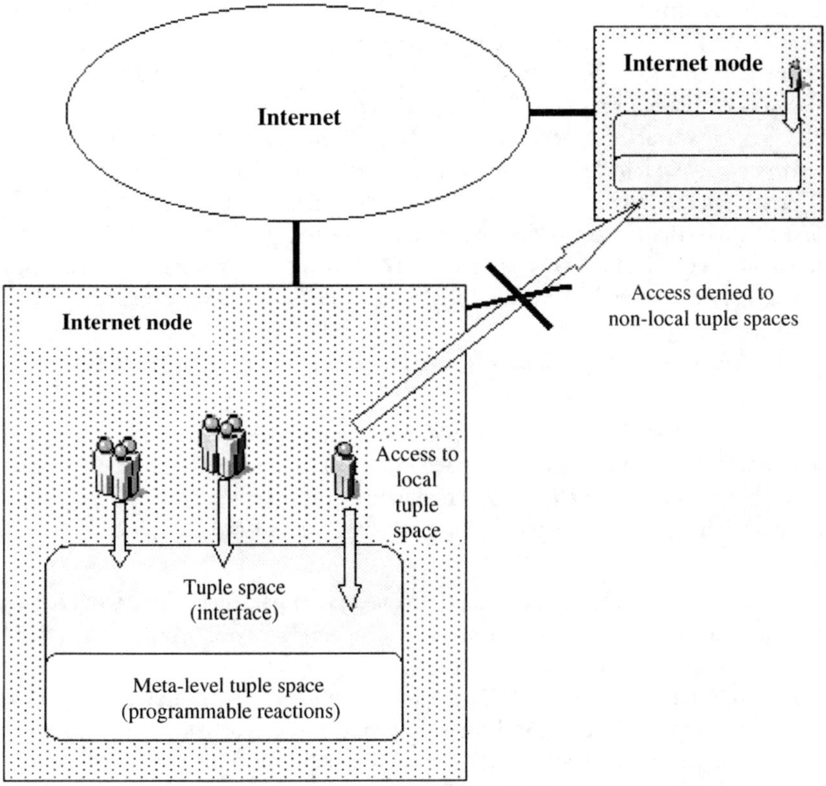

Figure 11: Mars architecture.

Furthermore, every Mars tuple space can react to accesses with a programmable behaviour defined in the so-called *meta-level tuple space*. Mars, in fact, defines a flexible and highly controllable architecture to plan the reactions of agents' accesses to tuple spaces. By using a four-field tuple (4-ple) in the form (Rct, t, Op, I), for example, Mars can make an Rct compatible reaction set off when the agent I invokes the operation Op on tuple T. A tuple's writing or taking in the meta-level tuple space by the network administrator or by authorized agents implies the installation or removal of particular reactions. A reaction can influence the effects of particular operations, since it is based on the current tuple space behaviour and on the previous accesses to it. It is easy to guess the importance of this model's advantages in terms of flexibility, control and design simplicity.

The aim of Mars project is not to build a new execution environment for mobile agents replacing current frameworks, but to develop a general and portable coordination architecture to place side by side to already available engines. Mars' tuple space is feebly connected to the agent server; therefore, it can be associated to different mobile agent systems, such as IBM Aglets or Odyssey engine among others.

5.5.1 Mars interface

Mars tuples are Java objects representing ordered sets of types, not necessarily elementary ones. In order to define a tuple, the *Entry* class (defining the basic features of tuples) must be extended and the specific fields of the tuple defined as instance variables. Every tuple's field can also represent primitive data types (*wrapper* objects, in Java terms), and can show both defined and null values.

Tuple space is carried out as Java object; it is precisely an instance of the *Space* class that implements the interface through which agents can access to tuple space. The adopted interface extends the one described in JavaSpace specifications, an architecture likely to become *de facto* the standard one in this field.

JavaSpace interface defines three operations in order to access to tuple space:

- *write*, to insert a tuple, given as parameter, into the space;
- *read*, to read a tuple from the space, on the basis of a tuple given as parameter and to be used as a pattern for the matching mechanism;
- *take*, working as read, except that it takes the tuple from the space.

In addition to tuple parameters, the *transaction* parameter gives details about the characteristics of the operation: It is used in Mars to specify if take or read operations are stopping ones or not: a non-stopping operation not finding any matching tuple goes back to NULL value. The *identity* parameter specifies who asks the operation and is used for safety reasons. Furthermore, Mars interface has got the two *readAll* and *takeAll* operations to extend the operations to all the matching tuples met in the research.

The tuple used as pattern for read and take operations can have both defined and null values: a tuple T will match the required tuple T if R's defined values are the same as T's corresponding ones. In particular, since in JavaSpace, as well as in Mars, tuples are objects and the elements of a tuple can be objects (not primitive ones like wrappers) too, matching rules must take into consideration such cases: a required tuple R matches a tuple T if and only if the following conditions occur:

- R is an instance of class T or of one of its super-classes; in this sense JavaSpace extends Linda model allowing also the matching between two different kinds of tuples, if they belong to the same hierarchy;
- R's defined fields representing primitive types (integer, character, boolean, etc.) have the same values as their corresponding fields in T;
- R's defined non-primitive fields (i.e. objects) are the same – in a serialized form – as the corresponding ones in T;
- a null value in T matches a null value in R.

Once a tuple has been taken from the space, its fields become accessible to any Java object.

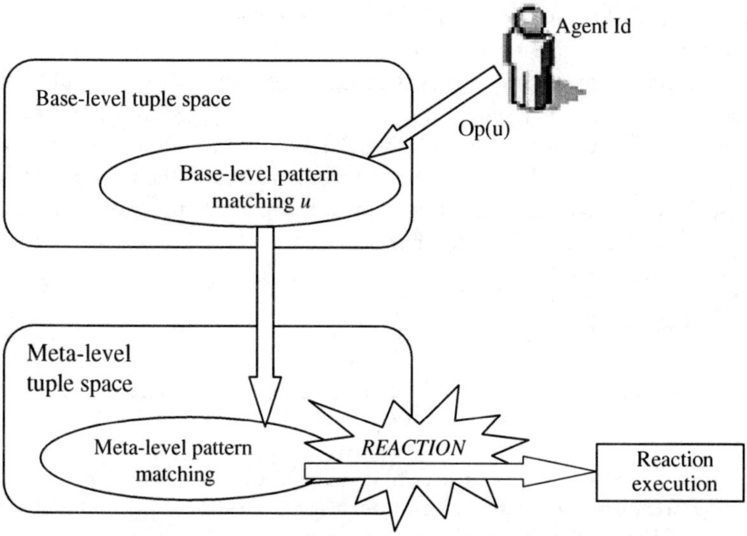

Figure 12: Meta-level tuple space's activity.

5.5.2 Reactive model

Differently from JavaSpace, Mars enables to associate programmable reactions to tuple space's events. Reactions are single-method objects that can have access to tuple space in their turn, change their content and influence the semantics of agents' operations.

There are four parameters associated to a reaction: the reaction (Rct), the tuple element (T), the type of operation (O) and the agent's identity (I); association is made therefore through a 4-ple (Rct, T, O, I) and the reaction, i.e. the method of Rct object, is executed when agent I invokes operation O on tuple T. When we refer to the association of reactions with tuples, we usually speak of *meta-level tuples* (fig. 12). The associated meta-level tuple space often follows the same mechanisms as the base tuple space: a 4-ple (ReactionObj, NULL, read, NULL) will specify to associate the reaction of ReactionObj to all the read operations, regardless of the particular kind of tuple and the agent's identity.

An aspect not yet investigated thoroughly is the one concerning the agent's identification and tuple spaces access permissions. JavaSpace specifications define a simple mechanism (based on access control list, ACL) to strengthen control to space accesses. In Mars, the ability of programming tuple space and the possibility of accessing and modifying its content through reactions introduces different safety problems. Obviously, a site's administrator must be able to program freely and without the constraints typical of tuple space; on the other hand, one can think of providing agents with a high degree of flexibility, leaving them the possibility of programming their meta-level spaces to install their own reactions, shifting the problem of safety to the choice of proper politics directly linked to the agent.

5.5.3 Context-dependent coordination in Mars

Mars system is suitable to describe also context-dependent coordination model: let us suppose, for example, that a mobile agent application should find web pages in a site, and that such pages are saved in htm format rather than in the classical html one. Instead of obliging the site's local administrator to change all extensions or double their tuples in order to make them available in both forms, the administrator may simply install a reaction transforming each html file request in an htm file one. The reaction will be installed by writing the meta-level tuple as follows:

```
class HTML2HTM implements Reactivity
{
  Public Entry[] reaction(Space s,Entry InputTuple[],Entry
                          Template,Operation Op,Identity Id)
  {
  //  if the site has just files with htm extension,
      modifies extension in the template
      ((FileTuple) Template).Extension="htm";
      // asks for matching with the new extension
      Entry[] result = s.readAll(Template,null,NO_WAIT);
      for (int i=0; i<result.length; i++)
      // restores extension in the found tuples
      ((FileTuple) result[i]).Extension = "html";
      return result;
  }     // end of method
}
```

As to environment-dependent coordination, a typical problem occurs when an agent tries to carry out a takeAll operation (i.e. an extraction from tuple space) of html files, thus eluding the administrator's control. The latter can, on the other hand, get round such disadvantage with a simple reaction transforming takeAll operation into a readAll one having the same parameters, thus returning the agent the data asked for in a completely transparent way. The reaction can be implemented as follows:

```
class ReadOnly implements Reactivity
{
      public Entry[] reaction(Space s,Entry InputTuples[],Entry
      Template,Operation Op, Identity id)
      {
       // controls the agent's identity
      if (Id.equals(manager))
          // the administrator can remove tuples
          return s.takeAll(Template,null,NO_WAIT);

          // otherwise tuples are not taken but simply
             returned
```

```
        else return InputTuples;
        } // end of reaction method
}
```

5.6 Models comparison

As we have already seen, all coordination models are characterized by three basic elements: coordinables, coordination media and coordination laws. To sum up, we can say that the main goals to focus on are:

- *Coordinables*: it is necessary to give a role to Internet services for a proper integration of the model in the Web. The only achieved attempt in this direction is PageSpace: every Internet service can find its role as *special-purpose coordinable*. Anyway, the idea of hiding Internet services behind a tuple space should not be excluded in advance, because it could be useful by providing a simple and tuple-based access to already existing services.
- *Coordination media*: coordination should be based on a variety of tuple-based and independent abstractions, distributed in Internet sites and possibly managed most suitably. A network-aware architecture undoubtedly implies some advantages, since mobile agents can implicitly refer to tuple spaces, according to their position. Moreover, they could be given the possibility of explicitly referring to a tuple space, for example thanks to a URL, as in TuCSoN. However, in the case of small-size networks (e.g. inner networks), transparency has the best advantages. Another interesting solution is the one presented by LIME: agents can transparently act according also to the other nodes' resources, as if they were local.
- *Coordination language*: a limited set of Linda-like primitives is often enough. Moreover, a coordination language should not be dynamically extensible, in order not to complicate application management. If new functionalities should be added, they can be incorporated as coordination laws, without modifying coordination language. Designing the *data model* is more difficult. The best solution would be to define an architecture where different tuple spaces, based on different *data models,* coexist.
- *Coordination laws*: while the simplicity of Linda-like coordination language perfectly suits Internet applications, Linda's coordination laws could be too limiting. That is why a programmable tuple space should integrate a *Web coordination architecture*. Systems must be able in fact to modify their behaviour in response to some particular communication events, without having to modify coordination language. From this point of view, programmable coordination media based on different models, such as the TuCSoN logical one or the object-oriented Mars one, should be able to coexist. It is clear that all suggestions should take into consideration current standards, as well as those that are gaining ground, in order to be more easily accepted. For example, since Xml technology will become the effective standard for data structure representation

and information exchange through the Internet, each model should consider that. To this purpose, in the following sections we will deal with Mars-X and XmlSpaces.

Another interesting problem is one concerning security. While a coordination language should enable other agents to exploit all communication primitives and to have access to all available data structures, any security policy would limit interaction protocol, for example by forbidding access to some resources or communication media. The best compromise between expressive power and interaction security must be looked for.

Finally, a coordination model should take into consideration economic problems, for example by fixing interactions prices, coping with limited resource situations, etc. We will see in the following how the different coordination models have interpreted such concepts, trying to understand their pros and cons [2, 25].

6 Definition of coordinables

Generally speaking, we cannot think to define coordinables without defining also non-mobile entities, which we can call non-mobile coordinables. This is a necessary distinction since an Internet application includes also various non-mobile entities, such as the WWW servers and the CORBA-compliant services [6]. The set of non-mobile entities can be considered as a general set of Internet services.

The usefulness of a Linda-like system making use of tuple spaces can be better understood if it is thought of as a coordination model embedded in a conventional programming language. In the case of Linda embedded in Java, for example, the coordinables could be active Java objects, the coordination medium a Java objects multiset, and the coordination laws could be those describing the semantics of Linda-like's primitives.

The tuple space can be used as a coordination medium between agents and Internet services. Fig. 13 shows a scheme of such situation.

It should be noted that an approach like that does not lead to an explicit and precise definition of coordinables. Such choice is adopted by Mars that associates a single global tuple space to each node in the network able to accept and execute agents. Mars gives an agent the possibility of interacting with the execution environment thanks only to tuple space. It enables the adoption of an extremely simple programming style, leading therefore to the definition of a likewise simple programming language, making Linda-like coordination the only possible way to interact with Internet applications. A strategy of this sort could however prove unsuitable for components, such as proxy-servers, that could need to act both as client and as server [26].

An alternative choice would be that of Web entities used as coordinables. PageSpace adopts this approach, shown in the scheme in fig. 14.

PageSpace is nowadays the only system defining and characterizing both coordinables (e.g. Internet agents) and all other entities forming the model. It

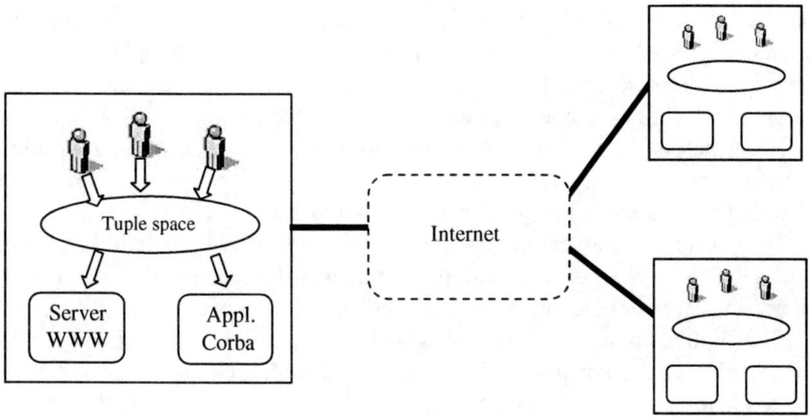

Figure 13: Tuple spaces as coordination media.

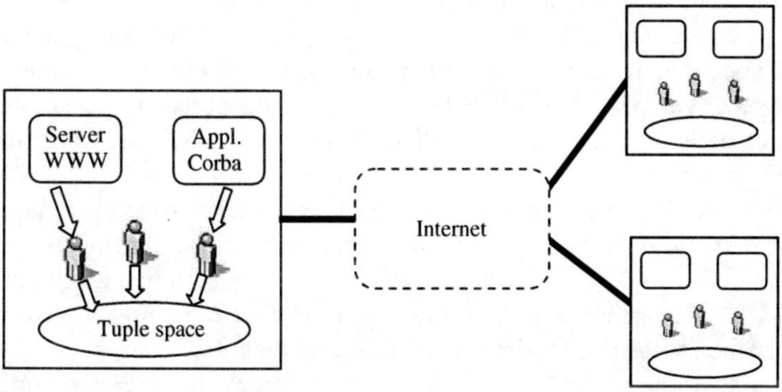

Figure 14: Tuple space as middleware.

is important to notice the difference from JavaSpace (and therefore Mars) and T Spaces, which define only coordination medium architecture, taking the presence of coordinables not better determined for granted.

6.1 Definition of coordination media

Even if simpler applications could benefit from the use of a single tuple space, it would not allow the development of complex and widely distributed applications. They need, in fact, several tuple spaces distributed all over the Internet and accessible through agents. The availability of multiple and independent tuple spaces, on the one hand, enables decentralization and modularity, on the other, introduces new problems. In particular, a coordination model including multiple tuple spaces needs that the agents have an effective way of relating each other and accessing the spaces. Many coordination systems (JavaSpace, Mars, PageSpace) suppose that tuple spaces themselves are objects. Agents can then exchange their references to tuple spaces and use them to access to a certain tuple space, transparently to their location. This extremely simple solution provides us a both transparent and location-independent way to access to tuple space.

Nevertheless, in the context of Internet applications, network unawareness is not the best choice, since unpredictable latency times typical of the Internet would turn the application performances unpredictable too.

Tuple spaces' organization, as well as their accessibility, must take into consideration the network-aware mobility of agents. From this point of view, two approaches are possible: an implicit one and an explicit one.

The implicit approach spares the agent the need of explicitly referring to tuple space and lets the agent automatically access a default tuple space varying according to its position in the Internet. Thus, implicit approach connects the coordination model questions to agents' mobility.

Mars system adopts an implicit and context-dependent approach to identify and have access to the tuple space. The tuple space can be used to have access to resources and services in a certain implicit way (without naming it or referring to it), as well as to enable coordination among agents. When an agent arrives on a new node, a new connection between the agent and that node's tuple space is established, while the old connection is lost. An agent has no way of accessing a remote tuple space but that of explicitly migrating to the node where that tuple space is situated.

LIME adopts a different approach: each agent carries with it a tuple space, and that is the only tuple space to which it can access during its existence. As soon as an agent arrives in a node, its private tuple space is automatically fused with the one associated to the execution environment. Thus, the private tuple space can be used to implicitly access to that node's resources. The main disadvantage of implicit approach is that it forces the agents to migrate in order to access to a tuple space. That evidently cannot be the best choice when the agent needs to access only once to the tuple space; in that case remote access would be more suitable. Explicit approach, by keeping a network-aware architecture, enables remote access to tuple spaces thanks to a *global naming scheme* (e.g. URL).

The adoption of an explicit approach does not imply the rejection of implicit approach: TuCSoN, for example, enables the agents to both implicit and explicit access.

6.2 Definition of coordination laws

In the context of Internet, object-oriented tuple space model adopted by JavaSpace takes the place of Linda's classical pattern-matching mechanism. Two objects match if their *serialized forms* are the same. Contrariwise, logical unification is taken as basic matching mechanism by TuCSoN. The majority of systems dealt with in the available literature on the subject (among which PageSpace, LIME and JavaSpace) adopt Linda-like approach that consists in including coordination fixed laws into the coordination medium. However, in a wide and unpredictable context as the Internet, this choice could not be the best one, since mobile agents must interact and migrate through heterogeneous execution environments.

T Spaces, TuCSoN and Mars see the coordination medium as a configurable kernel, so that new coordination laws can be defined, for the sake of flexibility. While T Spaces make it possible to define new primitives, TuCSoN and Mars forbid both the adoption of new ones and the modification of current matching mechanisms. They make it possible though to program anew coordination laws, thus enabling the definition of new response behaviours to communication events. In TuCSoN meta-level intelligent agents can program coordination media. Mars, consistently with its object-oriented tuple space, follows an object-oriented approach to program coordination laws: a specific method of a specific interface-class can be implemented and associated to a specific communication event. That makes Mars programmability desirable for high-level services and *system management.*

To conclude our remarks on coordination laws, we think it advisable to briefly deal with the question of coordination language, which we had previously defined as the ensemble of interaction primitives. The current trend, followed by PageSpace, JavaSpace and Mars, is that of defining tuple spaces object-oriented models, where both tuple spaces and components are Java objects. Nevertheless, Java could not be the most suitable language for all applications. For example T Spaces, used to manage large data, adopts a relational model: every tuple is a table's line; every kind of tuple identifies a kind of table.

As far as communication primitives are concerned, the majority of systems identify a limited set of primitives having the same semantics of Linda-like primitives. Asynchronous operations similar to JavaSpace notification mechanisms are useful for applications based on autonomous agents: an agent can ask a tuple to a tuple space without being stuck waiting for the right one; as soon as notification of right tuple availability comes, it can recover it. When mobile agents are involved, however, notification mechanisms are not convenient, since they would require run-time support to locate *mobile agents.*

7 Projects in progress

So far mobile agent technology has not imposed itself outside the boundaries of university research, not so much because of lack of maturity of available solutions, but for the lack of a standard clarifying the basic concepts on which an architecture must be based: coordination, communication, fault-tolerance, security, etc. For coordination and interoperability, the situation seems to be somewhat better, since in the last few years special groups, first of all the OMG consortium, are studying solutions thought of as standards open to all systems.

In this perspective, we can consider many of the projects concerning coordination models combined with interoperability mechanisms, first of all the promising use of Xml technology that is already having great success within the area of content exchange on the Internet.

As we have already widely seen, tuple space coordination models are an interesting technology facilitating the development of mobile agent distributed applications; moreover, the increasing number of projects making use of Linda-like coordination, dealing with it through different approaches, cannot but confirm such a trend. Anyway such methods must still be checked on actually used Internet applications. One of the points to be taken into consideration is security, intrinsically linked to coordination. While coordination technologies for distributed applications lead towards increasing interaction in order to make it more usable, security tends to limit those interactions, in order to have a more direct control. In other words, while coordination language tends to allow every agent to access any available resource, security policy tends to reduce and limit such accesses, hindering for example the access to some data or devices.

Current standards as well as other emerging ones are advancing in the area of mobile agents. For example since Xml technology is an actual standard for data representation and Internet exchange, it is easy to predict that every coordination model will become compatible with Xml. It is in fact possible to represent in Xml not only data, but also agents and services, as well as high-level interaction protocols (fig. 15). It does not mean that Xml will replace current coordination architectures, but these models, by integrating Xml technology, will lead to a framework where a wide range of agent-based proposals will find a coherent role.

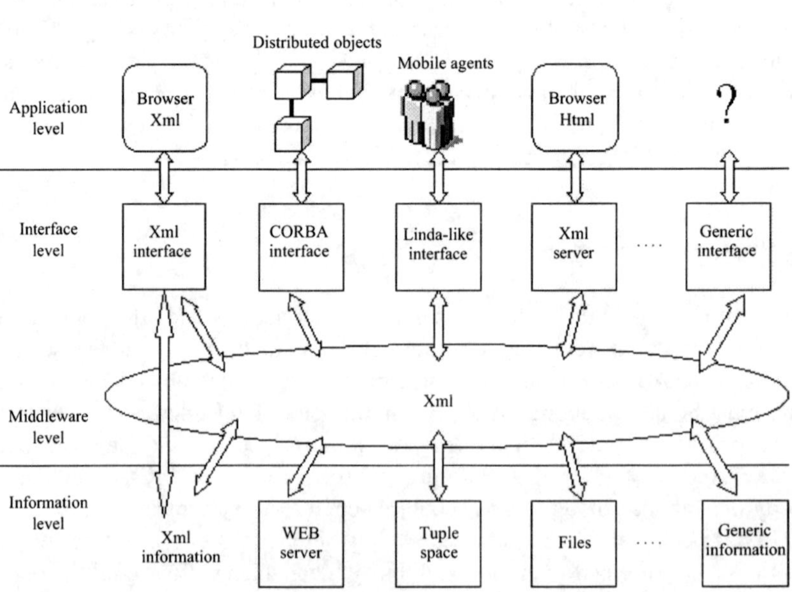

Figure 15: Xml architecture.

This section is framed within this field; it introduces Mars-X and Xml-Spaces, two integrated architectures for mobile agent coordination through Xml technology.

7.1 Mars-X

Mars-X was born from the combination of Xml and Mars [27]; it is a kind of coordination architecture for mobile agents combining the advantages of Xml language (interoperability among heterogeneous information sources) with Linda-like coordination (dynamism and completely uncoupled interactions). In Mars-X the agents coordinate with each other and also with execution environment, thanks to programmable Xml dataspaces. Each of them is associated to a particular execution environment and is Linda-like accessible to agents, as if it were a normal tuple space. Since the behaviour of Xml dataspaces in response to accesses can be totally programmed, Mars-X enables to include both application-specific and coordination-specific coordination rules in dataspaces.

In the case of Mars-X, the Linda-like interface enables the agents to access Xml dataspaces in a standard JavaSpace interface (fig. 16). It means that a Mars-X tuple space takes on the form of a Java object, which makes the operations read, write and take available, as well as the readAll and takeAll ones that Mars adds to JavaSpace.

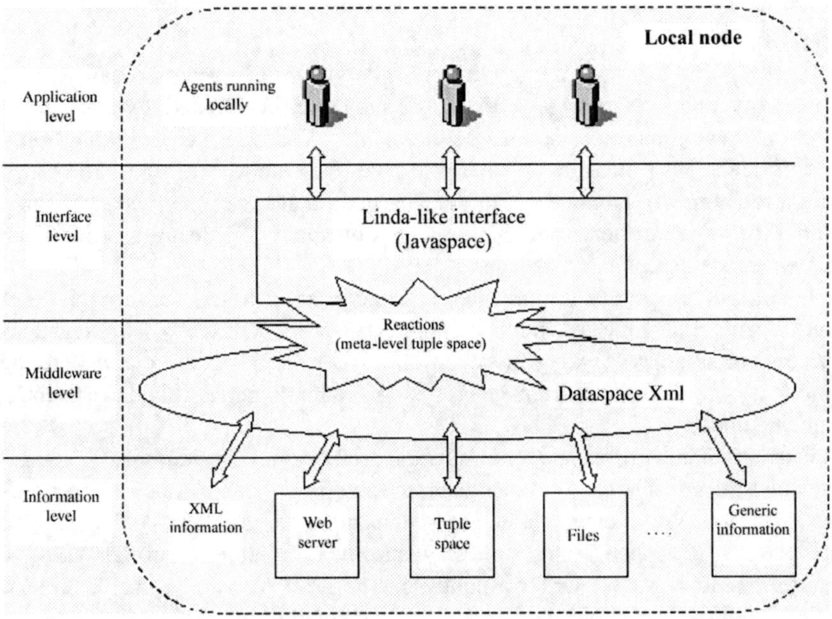

Figure 16: Mars-X architecture.

The choice of implementing JavaSpace interface rather than defining an Xml-oriented one has been dictated/imposed by the fact that, as we have already said, JavaSpace technology will be very important in the context of distributed Java applications in the nearest future. Furthermore, this choice best suits the approach of Java agents that see tuples as Java objects whose variables are their own fields. Interface operations must be able to translate tuples object representation into the corresponding Xml representation and *vice versa*.

In JavaSpace as well as in Mars-X and in Mars, Entry interface is implemented by deriving every tuple class from the AbstractEntry class. Furthermore, in Mars-X every tuple class must have an extra field specifying DTD (document type definition), which describes the structure of the Xml document matching that class's instances.

Every tuple of a generic tuple space matches one and only one element of an Xml document. In particular:

- the class' name, beginning with an underscore, matches, once the underscore is removed, the tag's name defining the Xml element;
- the instance variables' values match the data included in the tags;
- when an agent invokes an input operation, Mars-X searches in Xml dataspace an element in an Xml document so that;
- the document's DTD is the same as the one specified in the tuple's field;
- the tuple translated in Xml format matches at least one element in the document;
- the values in the tuple's field match the values of tags in the element.

In the point of view of mobility, in order to obtain access locality, Xml dataspaces must be considered as local resources associated to a node. In Mars-X, every Internet node must define its own Xml dataspace and the Linda-like interface associated to it. When it arrives on a node, an agent is provided with a reference to the node's Xml dataspace, so that it can freely access it in a Linda-like way. Furthermore, local domains of nodes (e.g. local networks) can implement a single Xml dataspace.

In Mars-X, reactivity is implemented exactly as in Mars, i.e. through reactions, events and 4-ple. It should be noted that Mars-X, exactly as Mars, does not implement a new Java agent system, but has been designed to complete the functions of the already existent agent systems, without being linked to a specific implementation.

Current Xml implementation suffers from some limitations: (a) tags' attributes are considered additional tags, so there is no "one-to-one" correspondence between Xml elements and Entry objects; (b) it does not deal with Xml namespaces; (c) concurrent accesses synchronization is done through MR/SW (multiple readers single writers) policy on the level of single Xml document, which is not the best choice.

7.2 XmlSpaces

Very similarly to Mars-X, XmlSpaces system [28], developed at Berlin's Technische Universität, extends the basic Linda-like model, in particular Linda's TSpaces implementation, in various ways:

- Xml documents are used as tuple fields in coordination space; thus, they can be dealt with as ordinary tuples as well as mere Xml documents seen as single-field tuples;
- a wide range of Xml document relations can be used for matching functions, some of which are naturally supported by the architecture while other extensions can be freely added;
- XmlSpaces is a distributed architecture, i.e. multiple servers in different locations used as dataspace are seen as a single logical dataspace; different security policies can be adopted simultaneously;
- distributed events are supported; they enable immediate notifications to all clients of a tuple's insert.

In XmlSpaces the fields tuple's *actual* fields can contain, as already said, entire Xml documents, while *formal* fields can contain further description about the document, such as a query expressed in a suitable Xml query language. The matching relation follows the same rules as Linda-like pattern matching, even if in this case multiple relations are supported on Xml documents. In particular, matching rules are based on XQL engine and take into consideration the following facts:

- an Xml document matches another one on the basis of their content equivalence or their attributes equivalence in the single elements;
- an Xml document matches another one on the basis of a minimum common grammar with or without equivalence of remaining elements and their attributes;
- an Xml document can be compared to a matching query following one of the supported semantics, such as Xml-QL, XQL or XPath/Pointer.

The use of XmlSpaces in the context of distributed applications, beyond the undeniable advantage of Linda-like language extension, finds its major strength in the use of different distributed schemes according to particular exigencies; actually, the architecture can be:

- *centralized*: a server encapsulates the whole dataspace;
- *distributed*: all servers encapsulate a subset of the whole dataspace;
- *full replication*: all servers possess persistent copies of the whole dataspace;
- *partial replication*: subsets of servers keep consistent copies of the subsets of the whole dataspace;
- *hashing*: matching tuples and templates are kept in the same server and selected according to some hash function.

In XmlSpaces, however, a specific use strategy is not prescribed, rather the distributed strategy is encapsulated in the so-called *distributor object*, to whose interface Linda-like primitives are provided. Thanks to the use of distributor object, XmlSpaces is completely open, since any server can join or leave at every moment, because the distributor object must simply record currently active servers through the record/delete methods included in its interface. Notification of distributed events to all the servers that have subscribed to the event in question will occur also through the distributor object, thus making the multicasting based on distributed Xml dataspaces easy to implement.

References

[1] Nwana, H.S., Lee, L.C. & Jennings, N.R., Co-ordination in software agent systems. *BT Technology Journal*, **14(4)**, pp. 79–88, 1996.

[2] Guaitoli, G., Infrastrutture di coordinazione dipendenti dal contesto in ambito wireless, http://polaris.ing.unimo.it/didattica/curriculum/letizia/tesi/guaitoli/Tesi.pdf, 2000.

[3] Cardelli, L. & Gordon, A.D., Mobile ambients. *Formal Methods for Distributed Processing, A Survey of Object-Oriented Approaches*, eds. H. Bowman & J. Derrick, Cambridge University Press: Cambridge, UK, ISBN 0-521-77184-6, pp. 198–229, 2001.

[4] Gelernter, D. & Carriero, N., Coordination languages and their significance. *Communications of the ACM*, **35(2)**, pp. 96–107, 1992.

[5] Reggiani, G., Cabri, G., Leonardi, L. & Zambonelli, F., *Design and Implementation of a Programmable Coordination Architecture for Mobile Agents, Technology of Object-Oriented Languages and Systems*, IEEE, 1999.

[6] Baumann, J., Hohl, F. & Straßer, M., Beyond Java: Merging Corbabased mobile agents and WWW, A position paper for the Joint W3C/OMG Workshop on Distributed Objects and Mobile Code, 1996.

[7] Cabri, G., Leonardi, L. & Zambonelli, F., Reactive tuple spaces for mobile agent coordination. *Proc. of the 2nd International Workshop on Mobile Agents,* number 1477 in LNCS, pp. 237–248, 1998.

[8] Ciancarini, P., Agent technology, http://www.cs.unibo.it/~cianca, 2000.

[9] Cabri, G., Leonardi, L. & Zambonelli, F., Coordination infrastructures for mobile agents. *Elsevier Microprocessors and Microsystems*, **25**, pp. 85–92, 2001.

[10] Cabri, G., Leonardi, L. & Zambonelli, F., How to coordinate Internet applications based on mobile agents, Seventh IEEE International Workshops on Enabling Technologies: Infrastructure for Collaborative Enterprises, (WET ICE 1998), pp. 104–109, 1998.

[11] Cabri, G., Leonardi, L. & Zambonelli, F., The impact of the coordination model in the design of mobile agent applications, IEEE Computer Software and Applications Conference, 1998.

[12] Lee, L., Nwana, H.S. & Jennings, N.R., Co-ordination in multi-agent systems. *Software Agents and Soft Computing, Towards Enhancing Machine Intelligence, Concepts and Applications*, eds. H.S. Nwana & N. Azarmi, number 1198 in LNCS, Springer-Verlag, London, UK, 1997.

[13] Bussmann, S. & Muller, J., A negotiation framework for co-operating agents. *Proc. of CKBS-SIG*, University of Keele, 1992.

[14] Khunboa, C. & Simon, R., On the performance of coordination spaces for distributed agent systems. *Proc. of 34th Annual Simulation Symposium*, Seattle, WA, USA, pp. 7–14.

[15] Denti, E., Natali, A. & Omicini, A., On the expressive power of a language for programmable coordination media, ACM Symposium on Applied Computing, Atlanta, 1998.

[16] Cabri, G., Leonardi, L. & Zambonelli, F., Engineering mobile-agent applications via context-dependent coordination. *Proc. of the 23rd International Conference on Software Engineering*, IEEE, 2001.

[17] Tolksdorf, R., Coordinating Java agents with multiple coordination language on the Berlinda platform, Sixth IEEE Workshops on Enabling Technologies: Infrastructure for Collaborative Enterprises, 1997.

[18] Lange, D.B. & Chang, D.T., IBM aglets workbench – Programming mobile agents in Java, IBM Corporation White Paper, 1996.

[19] Peine, H., Application and programming experience with the Ara mobile agent system, Preprint of an article accepted for publication in IEEE Software – Practice and Experience, 2002.

[20] Lingnau, A., ffMAIN: Using Tcl and the TclHttpd Web Server to implement a mobile agent infrastructure, www.tu-harburg.de/skf/tcltk/papers2000/paper.pdf, 2000.

[21] Sun Microsystems, The JavaSpace specification, http://www.sun.com/jini/specs/js-spec.html, 1999.

[22] The MOON Home Page, University of Modena, http://sirio.dsi.unimo.it/MOON.

[23] Ciancarini, P., Tolksdorf, R., Vitali, F., Rossi, D. & Knoche, A., Coordinating multi-agents applications on the WWW: A reference architecture. *Software Engineering, IEEE Transactions*, **24(5)**, pp. 362–375, 1998.

[24] Cabri, G., Leonardi, L. & Zambonelli, F., Mobile-agent coordination models for Internet applications, *IEEE Computer*, **33(2)**, 2000.

[25] Ciancarini, P., Omicidi, A. & Zambonelli, F., Coordination models for multi-agent systems, http://www.agentlink.org newsletter 3, 1999.

[26] Ciancarini, P., Omicini, A. & Zambonelli, F., Coordination technologies for Internet agents. *Nordic Journal of Computer*, **6**, 1999.

[27] Cabri, G., Leonardi, L. & Zambonelli, F. Xml dataspaces for mobile agent coordination, ACM 1-58113-239-5/00/003, 2000.

[28] Tolksdorf, R. & Glaubitz, D., XmlSpaces for coordination in web-based systems, flp.cs.tu-berlin.de/~tolk/xmlspaces, 2001.

Interoperability

Marco Di Stefano, Salvatore Vazzano and Alessandro Genco
DINFO – Dipartimento di Ingegneria Informatica

1 Introduction

Interoperability among agents is one of the requisites for a large-scale development and spreading of mobile agent applications. Interoperability gathers all the devices able to make communication possible among network agents [1] and their platforms, regardless of factors such as the characteristics of the starting system, the language codifying them and the aim for which they have been entered in the network.

The attempts at achieving a communication standard are various; some of them are having great success. Among them we can remember MASIF [2, 3], whose specifications [4, 5] have been produced by object management group (OMG) [1], based on the now solid CORBA technology [1, 6, 7, 8] enabling any kind of objects to interact, using a communication channel called object request broker (ORB) [2]. OMG, moreover, has defined interface definition language (IDL) [9, 10, 11, 12] that enables to define objects interfaces, giving designers the possibility of working with any programming language supporting IDL mapping.

Among them we can mention Java, which for its portability characteristics well combines with distributed technologies and, for that reason, has had great success in the area of mobile agents [10].

Interoperability among different kinds of agent systems is feasible if operations like an agent's transfer or management are standardized. OMG has worked in this direction, producing MASIF (mobile agent standard interoperability facilities) specifications in November 1995. They do not define, however, a standard on certain "local" operations such as interpretation, serialization, execution and deserialization of an agent, since they are operations that, even if differently implemented in each platform, do not influence their interoperability.

Next to MASIF, there is FIPA (Foundation for Intelligent Physical Agent), an international organization, whose birth was due to the collaboration among various companies and universities. It deals with the development of technologies based on intelligent agents and provides specifications for the achievement of platforms able to make agents communicate with each other.

One of the interesting things about FIPA is the possibility of overtly using its technology. It is in fact going in the direction of creating a standard that is both accessible and usable by everyone without any kind of profit.

One of the first goals of FIPA was that of gathering a wide approval among the various organizations interested in agent technologies, in order to start standardization process. At present it has more than 56 members and organizations in 17 countries, including *Alcatel, BT, NTT, IBM, HP, France Telecom, Siemens, Hitachi* and *Toshiba.*

2 CORBA

CORBA (common object request broker architecture) technology was born in the autumn of 1991 as the fruit of research made by the OMG to define an infrastructure ensuring a standard in communication among network distributed remote objects, independently of the language used to codify them and of execution platform.

CORBA actually consists of a software layer (middleware) enabling to consider the network as a single channel where to send objects; that channel is called ORB. Communication among all CORBA components is always mediated and managed by ORB.

CORBA remarkably simplifies the achievement of distributed applications because it not only enables designers to disregard programming problems concerning platform heterogeneity and object distribution, but also provides a series of resources and permanent components that can be used at any moment from any Internet site without having to know their "physical" position.

Such an ability of invoking components whose location is unknown to client is called *location transparency*; it is one of CORBA's strong points.

It is ORB in fact that looks for the invoked objects and does it in a completely transparent way for client, thus relieving the programmer of problems such as the search for various components in the network and the entry of its own ones into the communication channel; in other words, ORB acts as a broker in remote objects communication.

The tool enabling to interface with such components is the IDL, a descriptive language for the implementation of object interfaces defined in CORBA specifications. Any programming language for which an IDL mapping has been defined can be used to implement CORBA objects; it will be in fact ORB to make translations into the right implementation language.

2.1 CORBA architecture

Let us analyse fig. 1 that outlines CORBA architecture [6, 7] and then show its components:

a. *Object* is an entity made of identity, interface and implementation.
b. *Servant* is an invoked class that implements remote object.
c. *Client* is an entity invoking the servant's methods producing a particular object entered into the ORB in the form of a request.

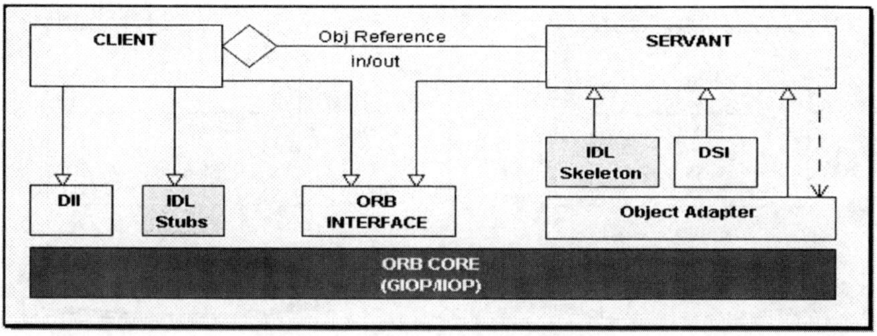

Figure 1: CORBA architecture.

d. *ORB* is a logical entity enabling to send the requests from a client delivering them to the remote object, in a complete transparent way; around it the whole CORBA technology develops.
e. *ORB interface* is the interface defining the ORB and enabling to hide implementation details; several interfaces can be defined for each ORB as long as an ORB-to-ORB communication protocol such as IIOP (Internet inter-ORB protocol that, by implementing TCP/IP communication protocol, is considered the standard CORBA) and GIOP (general inter-ORB protocol) are used;
f. *IDL stub* and *skeleton* operate in turn as cement between client and ORB and between ORB and server; both are automatically generated from the IDL definition of the interface associated to the object;
g. *DII (dynamic invocation interface)* is the interface enabling a client to dynamically send a request to a remote object, without knowing its interface definition and without having any bond/connection with the stub;
h. *DSI (dynamic skeleton interface)* is the analogous of DII for the Server, since it enables an ORB to deliver a request to an object that has not got a static skeleton;
i. *Object adapter* supports the ORB in delivering the request to an object as well as in object activation/deactivation operations.

2.2 The invocation of a remote object

Every interaction among distributed objects needs a client and a server; the latter has a remote interface that can be invoked by a client. It is also possible to have clients as objects, since they have remote object interfaces, having at the same time interfaces that can be invoked by other greater objects.

The client keeps the reference to the remote object; this is operated by a stub method, enabling to enter the object into the ORB before server invocation.

The server, on the other hand, uses the skeleton to transform the remote invocation into the local object call method (fig. 2). The skeleton, therefore, is in charge of transferring the invocation and its parameters into the corresponding

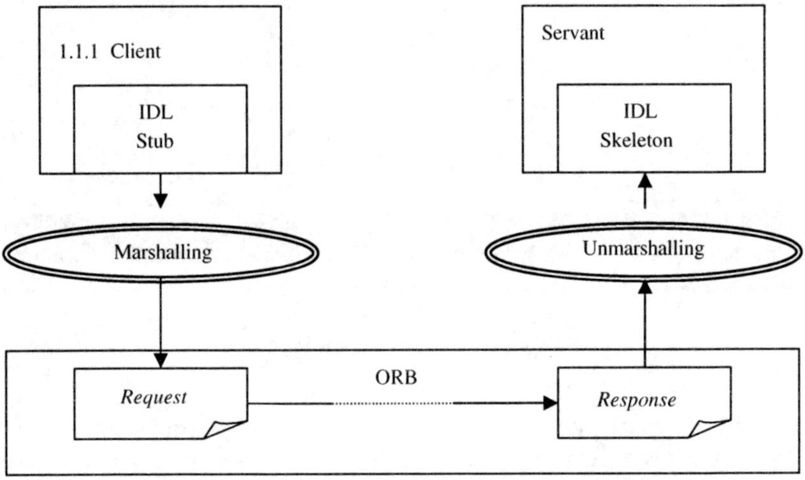

Figure 2: A remote invocation scheme.

implementation format, thus calling the invoked method. When the method stops, the skeleton enables the result (sometimes an error code) to travel through the ORB to eventually come back to the client.

As already said before, the communication protocol defining how information travels around the ORB is the IIOP that implements the TCP/IP and, like ORB and IDL, has been defined by OMG.

ORB also provides a great number of services for object management, such as object search by name, mobile agent system support, maintenance of persistent objects and many others.

CORBA object research systems are particularly interesting. The most common one is *naming service.* It enables to search for an object in the network, specifying its name, and obtaining a reference of it. Another method is the *StringField Object Reference* that is very useful when it is not possible to use a naming service.

2.3 Interface definition language (IDL)

As we have already hinted at, every object needs a particular interface detailed through an IDL definition; such interface is the basic requisite in order to make use of an object through CORBA and enables a client to interact with any object whose interface is known.

IDL is a language with syntax similar to that of C and Java, with a few syntactic differences (fig. 3).

When a CORBA object must be created, its interface must first of all be defined. It must be put into a file with .idl extension and given to the ORB's compiler.

Mapping	*CORBA version*
Ada	CORBA 2.0
C	CORBA 2.1
C++	CORBA 2.3
COBOL	CORBA 2.1
JavaToIDL, IDLToJava	CORBA 2.3
Smaltalk	CORBA 2.0

Figure 3: Languages supporting mapping through IDL.

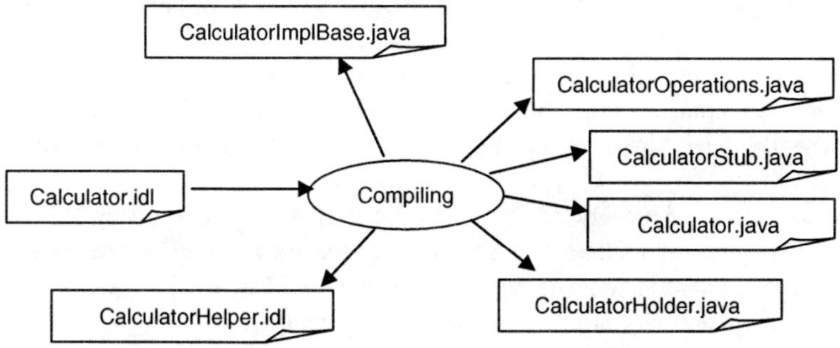

Figure 4: Compiling an IDL file.

Let us suppose we have defined the implementation for a *Calculator* object contained inside the *Calculator.idl* file and we want to use a Java compiler.

Through command line (*idlj* or *idltojava*) the Calculator.idl file is compiled whose result is the creation of a Calculator subdirectory containing a series of .java extension files (fig. 4):

CalculatorImplBase.java: it is the abstract class defining the skeleton; it is therefore in charge of providing the reception mechanisms of a request by the ORB and the replying back ones; the servant class implements it.

CalculatorStub.java: it is the abstract class defining the stub; it contains therefore the remote object conversion mechanisms from method invocation to invocation via ORB.

CalculatorOperations.java: it is the Java interface containing the object methods; it is used in server-side mapping as well as a mechanism for the optimization of client and server calls.

Calculator.java: it is the abstract class containing the Java version of the specified IDL interface. It implements the org.omg.CORBA.Object interface.

<UserType>Helper.java: for each type defined by the user a *Helper* file is generated; it contains some static methods implementing various functions to manipulate the associated type; among them we must remember the narrow method enabling to make a CORBA object cast.

CalculatorHolder.java: it holds the instance of each member of the class. It provides some operations for the management of *out* and *inout* parameters (which we will deal with later), present in IDL but not easily mapped in Java.

2.4 IDL syntax and Java mapping

IDL proposes quite a wide range of basic types [11], since it tries to keep as far compatibility as possible with a great number of languages. Thus, we will find the boolean type that does not exist in Java and is mapped literally through *true* and *false*; we will find the *wchar* and *wstring* types supporting 16 bits Unicode characters (for Java compatibility), but also *char* and *string* types dealing with 8 bits ASCII characters.

In order to define a Java package the keyword *module* is used (fig. 5).

Within a *module* it is possible to define one or more *interfaces*, which enable to specify the object interface by defining its *attributes* and *operations* (features and methods).

The interface of a CORBA definition is mapped by generating two files: <name_file>.java and <name_file>Operations.java. Attributes are meant as the interface's member data that can be simple or structured. There are three categories of structured data: *enum*, *union* and *struct*.

Enum is an ordered list of identifiers, such as:

```
enum PeopleNames {Paul, Liza, Marc, Vera};
```

It is mapped in Java through a class having the same name that implements the IDLEntity interface, with some useful methods for its management (among which a constructor and a function for the integer conversion).

Union is a combination of a "Union" and a "switch" instruction from C language:

```
union PersonalQuestions switch (PeopleNames) {
    case Paul: string favouritecolour;
    case Lisa: short eta;
    case Marc: boolean ready;
    case Vera: PeopleNames BestFriend;
};
```

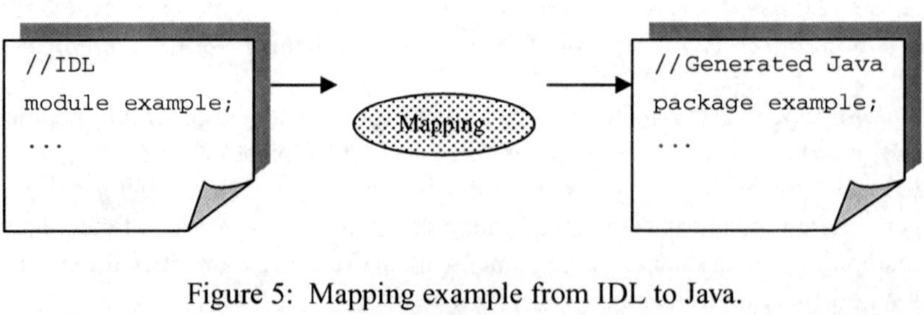

Figure 5: Mapping example from IDL to Java.

Union is mapped in a final class having the same name; it also implements IDLEntity interface. Moreover, the class is provided with some useful methods for its management.

Struct is a complex data structure that can contain various fields:

```
struct AtFriendsHouse {
    short NumFriends;
    PeopleNames Guest;
    string Host;
};
```

Struct is also mapped by generating a final class with the same name implementing IDLEntity interface and containing a default constructor method.

Actions are the interface methods. As far as methods are concerned, it should be noted that differently from Java, where the kind of parameters passage (by value or reference) was implicitly associated to type, in IDL the kind of passage one wants to adopt must be declared, which can be done by using the keywords *in*, *out* or *inout*.

A parameter *in* is passed by value, a parameter *out* by reference, a parameter *inout* in both ways.

Two other kinds of data exist: *sequence* and *array*; they are both mapped in Java with the *array* basic type. It must be added, however, that the sequence type can have a predefined dimension (bounded) or a non-predefined one (unbounded), differently from the array one.

The syntax to define a sequence unbounded datum is the following:

```
typedef sequence<int> IntSequence;
```

As far as constants are concerned, it is possible to define them both inside and outside the interface; in the first case constant will be generated inside the interface where it is, and in the second case it will be put inside an interface suitably created to contain it, with the same name.

Particular attention must be given to the *exceptions* that in IDL are treated in a very similar way to structs. The method that can call an exception uses the keyword *raises* (`throws` in Java). Let us see an example of IDL definition implementing an exception:

```
//IDL
module RaiseException{

 exception MyException{
 string cause;
 short ID_Exception;
 };

 interface RaiseException{
 string ICauseAnException() raises (MyException);
 };
};
```

If this simple .idl file is compiled, the subfile RaiseException will be generated containing the files:

```
_RaiseExceptionStub.java,
RaiseExceptionHelper.java,
RaiseExceptionHolder.java,
RaiseException.java,
RaiseExceptionOperations.java,
MyException.java,
MyExceptionHelper.java,
MyExceptionHolder.java.
```

The last three classes are abstract classes needed for exception management. The mapping between the IDL constructs and the Java ones is summarized in table 1.

Table 1: Mapping between the IDL constructions and the Java ones.

IDL construction	Java construction
module	package
interface (non-abstract)	signature interface and an operations interface, helper class, holder class
interface (abstract)	signature interface, helper class, holder class
constant	public interface
boolean	boolean
char, wchar	char
octet	byte
string, wstring	java.lang.String
short, unsigned short	short
long, unsigned long	int
long long, unsigned long long	long
float	float
double	double
fixed	java.math.BigDecimal
enum, struct, union	class
sequence, array	array
exception	class
readonly attribute	accessor method
readwrite attribute	accessor and modifier methods
operation	method

2.5 CORBA and mobile agents

Though seemingly they are similar, CORBA and MA technologies have some different features.

CORBA takes for granted that an object is always in the same host, and that, once its reference hooked up, it can be invoked in a remote way; on the contrary, mobile agents do not have a fixed location, and there is no guarantee that once found, they do not move to another location, losing thus their reference. It is therefore necessary to introduce new search and, above all, "maintenance" methodologies for such reference.

Another important difference is that mobile agents, migrating from one site to another, are always aware of the location they are in and know where to go to find the resources they need, whereas CORBA tends to hide the location of a particular component whose implementation has been invoked to the programmer, thus making almost impossible to implement mobile agents mobility features.

Finally, the spreading of both technologies over the network must also be taken into consideration; CORBA has already been accepted as a standard and an increasing number of accessible and usable resources and components can be found in the network, whereas, as far as mobile agent technology is concerned, a standard has not been defined yet, so there exist hundreds of different engines suffering from the impossibility of mutual communication, which undoubtedly discourages from a large-scale expansion of mobile agents.

A possible fusion of CORBA and MA should, therefore, solve all those problems and give birth to a new technology having the advantages of both, getting thus into the gear for the development of mobile agents.

3 OMG MASIF

In order to make interaction among different platforms possible, MASIF proposes a standard for the following operations:

a. operations for the *management* of agents, such as the creation, termination, suspension and following restarting of their execution;
b. operations for the *transfer* of agents from one agent system to another;
c. operations for the *identification* of agents, agent system and place through a string containing the following fields: *authority* that defines the person or organization for whom the entity operates, *identity* that univocally identifies a particular agent instance, agent system or place within the same authority, *agent system type* that identifies the kind of agent system it refers to;
d. operations for the *localization* of agents, agent system and place. Inside each region there must be a mechanism enabling the recording of all entities belonging to the region itself, so that they are accessible, everywhere in the region, in a transparent way to their physical location (fig. 6).

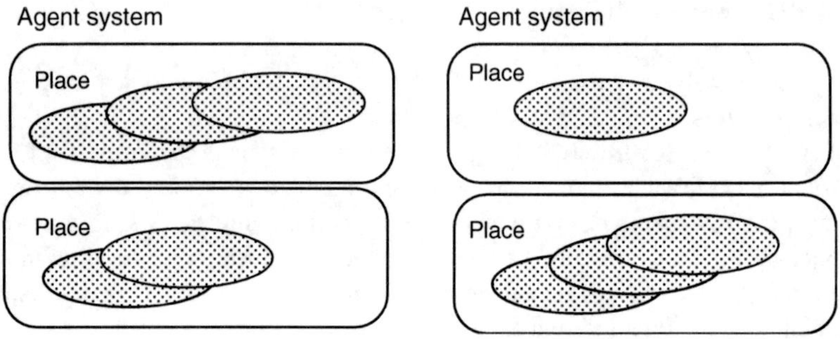

Figure 6: Architecture of a region.

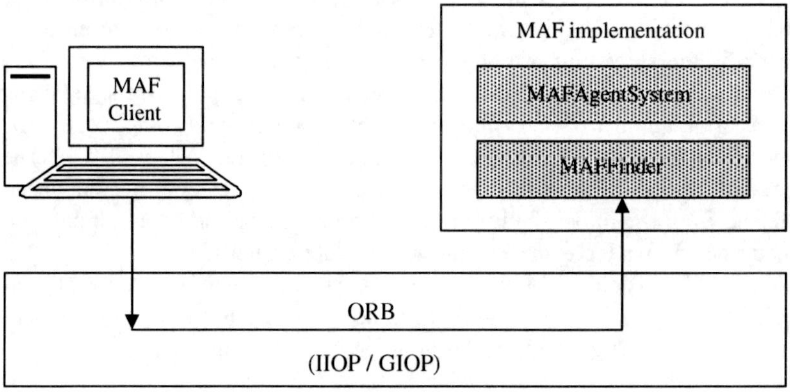

Figure 7: Connection between MASIF and ORB.

Functionalities are specified by defining two particular interfaces in IDL (fig. 7):

- *MAFAgentSystem*: it deals with agents management, transfer and execution operations; such interface must be implemented by any platform to enable interaction with agents coming from different platforms.
- *MAFFinder*: it is a naming service dealing with agents and agent system recording/unrecording operations and enables their identification.

At present MASIF provides the functionalities required for the first level of interoperability, consisting in the transport of agents from an agent system to another. Once the information concerning the agent has been transferred, the way the destination agent system manages such information depends on the implementation of the agent system itself and is not treated according to MASIF standard. It does not standardize in fact any operation for the local management of agents, such as those concerning their interpretation, serialization/deserialization and execution. In MASIF interfaces are defined at the agent system level rather than at the agent level.

3.1 IDL specification in MASIF protocol

IDL specification in MASIF protocol is a set of definitions and interfaces defining a communication standard for mobile agents.

Defined interfaces must be implemented at the agent system level: no interface is, on the contrary, given for the agent because being an agent's life confined inside an agent system, its implementation depends on the one of the agent system that has created it and from whom it inherits interoperability capacities.

Being moreover clear that an agent can migrate only from an agent system to another one supporting the same *agent profile*, there is no need of standardizing the agent's interfaces. Operations such as suspension, reactivation and killing of an agent must be standardized, as they provide a set of basic operations for agent migration management.

3.1.1 Name, Class Name and Location

Name structure is composed of three fields: *authority*, *identity* and *agent_system_type*. These attributes enable the association of a single identifier to each agent system.

When the *Name* defines an agent, the attribute *agent_system_type* indicates the type of agent system that has generated its identifier; on the contrary, when it defines an agent system, the attribute *agent_system_type* specifies the type of agent system.

The attribute *authority* defines the person or organization the agent or the agent system represents.

Finally, the attribute *identity* is an identifier generated through some mechanisms that can be different in various agent systems.

The combination of such attributes ensures us that, in any case, a single identifier will be created enabling us to recognize our agent or agent system.

```
typedef short AgentSystemType;
typedef sequence<octet> OctetString;
//The structure enabling the unambiguous
  identification of a class
struct ClassName {
  string name;
OctetString discriminator;
};
typedef sequence<ClassName> ClassNameList;
typedef sequence<OctetString> OctetStrings;
typedef OctetString Authority;
typedef OctetString Identity;

//The structure defining the name of an
  agent or agent system
struct Name {
  Authority authority;
  Identity identity;
```

```
  AgentSystemType agent_system_type;
};
//String containing the URL address or the URI
  name of an agent
typedef string Location;
typedef sequence<Name> NameList;
typedef sequence<octet> Arguments;
typedef sequence<Location> Locations;
enum AgentStatus { CfMAFRunning, CfMAFSuspended,
  CfMAFTerminated };
```

The structure *Class Name* is an identifier for classes that are accepted by an agent system. It is composed of a *name*, which must be readable and that can be interpreted by man, and a *discriminator* ensuring its unicity inside the execution context (fig. 8).

The way *name* and *discriminator* are generated must be defined by the agent system that must make sure of generating single names inside the place where the agents are received.

Location is used in MAFFinder interface and is used to specify the location of an agent system that is being searched for. For example, if we use an agent research method through the identifier Name, a location specifying the agent system's name containing it is returned. The location string can have two forms: it can specify an URI (containing a CORBA name) or an URL (containing an Internet address).

3.1.2 Authority identification

Specifications do not give details on what type of agent system, language, serialization mechanism and authentication methods must be used in order to host new agents.

Such characteristics must therefore be specified inside the AgentProfile structure; for example, we could set "Java" for language_id, "Aglets" for agent_system_type and "Java Object Serialization" for serialization.

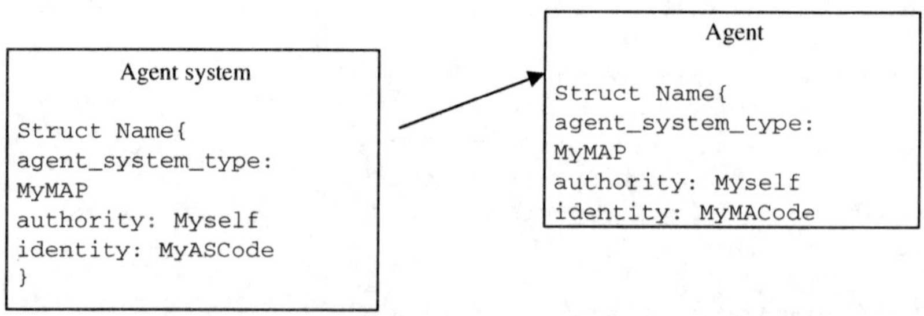

Figure 8: Name structure for an agent and agent system.

AgentSystemInfo structure defines some basic properties for the agent system, while the AgentProfile identifies an incoming agent.

```
typedef short LanguageID;
typedef short Authenticator;
typedef short SerializationID;
typedef sequence<SerializationID>
 SerializationIDList;
typedef any Property;
typedef sequence<Property> PropertyList;

//Map of languages that the agent system can
 accept
struct LanguageMap {
 LanguageID language_id;
 SerializationIDList serializations;
};
typedef sequence<LanguageMap> LanguageMapList;

//Class containing information on the agent system
 struct AgentSystemInfo {
Name agent_system_name;
AgentSystemType agent_system_type;
 LanguageMapList language_maps;
string agent_system_description;
 short major_version;
 short minor_version;
 PropertyList properties;
};

//Structure containing information on
 authentication
struct AuthInfo {
boolean is_authenticated;
 Authenticator authenticator;
};
//Structure defining the profile of an incoming
 agent
struct AgentProfile {
 LanguageID language_id;
 AgentSystemType agent_system_type;
 string agent_system_description;
 short major_version;
 short minor_version;
 SerializationID serialization;
 PropertyList properties;
};
```

Table 2: Table of exceptions of MASIF specification.

Exceptions	Meaning
`Exception AgentNotFound {};`	The agent does not exist
`Exception AgentIsRunning {};`	The agent is already running
`Exception AgentIsSuspended {};`	The agent is already suspended
`Exception ArgumentInvalid {};`	Arguments are not valid
`Exception ClassUnknown {};`	Error in class searching
`Exception DeserializationFailed {};`	Error in agent's decoding
`exception EntryNotFound {};`	The agent is not in the system
`exception FinderNotFound {};`	MAFFinder could not be found
`exception MAFExtendedException {};`	Unknown error
`exception NameInvalid {};`	An invalid name has been given
`exception ResumeFailed {};`	The agent could not be resumed
`exception SuspendFailed {};`	The agent could not be suspended
`exception TerminateFailed {};`	The agent could not be terminated

3.1.3 Exceptions

Table 2 shows a list of possible exceptions in the MASIF specification and their meanings.

3.1.4 MAFAgentSystem interface

```
interface MAFAgentSystem {

  Name create_agent(
   in Name agent_name,
   in AgentProfile agent_profile,
   in OctetString agent,
   in string place_name,
   in Arguments arguments,
   in ClassNameList class_names,
   in string code_base,
   in MAFAgentSystem class_provider)
  raises(ClassUnknown, ArgumentInvalid,
  DeserializationFailed, MAFExtendedException);
 OctetString fetch_class(
   in ClassNameList class_name_list,
   in string code_base,
   in AgentProfile agent_profile)
  raises(ClassUnknown, MAFExtendedException);
```

```
Location find_nearby_agent_system_of_profile(in
AgentProfile profile)
  raises(EntryNotFound);

AgentStatus get_agent_status (in Name
agent_name) raises(AgentNotFound);

AgentSystemInfo get_agent_system_info();

AuthInfo get_auth_info(in Name agent_name)
raises(AgentNotFound);

MAFFinder get_MAFFinder()
raises(FinderNotFound);

NameList list_all_agents();

NameList list_all_agents_of_authority(in
Authority authority);

Locations list_all_places();

void receive_agent(in Name agent_name, in
AgentProfile agent_profile, in OctetString
agent, in string place_name,   in ClassNameList
class_names, in string code_base, in
MAFAgentSystem agents_sender)
raises(ClassUnknown, DeserializationFailed,
MAFExtendedException);

void resume_agent(in Name agent_name)
raises(AgentNotFound,
 ResumeFailed, AgentIsRunning);

void suspend_agent(in Name agent_name)
raises(AgentNotFound, SuspendFailed,
AgentIsSuspended);

void terminate_agent(in Name agent_name)
raises(AgentNotFound, TerminateFailed);

void terminate_agent_system()
  raises(TerminateFailed);
};
```

Hereafter all the interface methods are listed:

- *create_agent*(): enables the creation of an agent satisfying a client requests and gives back the created agent's name. If the client is responsible for the naming, the name will be the same as the parameter agent_name given for its creation.
- *fetch_class*(): this method returns one or more classes definitions.
- *find_nearby_agent_system_of_profile*(): this method asks the MAFFinder to find the nearest agent system that can execute the agent sent by the client.
- *get_agent_status*(): it returns the specified agent state (running, suspended, terminated).
- *get_agent_system_info*(): this method returns the AgentSystemInfo structure, containing information about the agent system.
- *get_authinfo*(): this method returns information about how an agent has been authenticated, and which method has been used.
- *get_MAFFinder*(): this method returns a reference to MAFFinder to locate agents, place and agent system.
- *list_all_agents*(): this method returns the list of all agents registered in the system.
- *list_all_agents_of_authority*(): this method returns the list of all the agents recorded in the system under the same authority.
- *list_all_places*(): this method returns the list of all the places present in the system.
- *receive_agent*(): this method is used to receive and instantiate an agent.
- *resume_agent*(): this method is used to resume the execution of a specific agent.
- *suspend_agent*(): this method suspends the execution of an agent.
- *terminate_agent*(): this method terminates the execution of an agent.
- *terminate_agent_system*(): this method terminates the execution of the agent system.

3.1.5 MAFFinder interface

```
interface MAFFinder {

  void register_agent(
    in Name agent_name,
    in Location agent_location,
    in AgentProfile agent_profile)
  raises(NameInvalid);

  void register_agent_system(
    in Name agent_system_name,
    in Location agent_system_location,
    in AgentSystemInfo agent_system_info)
  raises(NameInvalid);
```

```
  void register_place(
   in string place_name,
   in Location place_location)
  raises(NameInvalid);

  Locations lookup_agent(
   in Name agent_name,
   in AgentProfile agent_profile)
  raises(EntryNotFound);

  Locations lookup_agent_system(
   in Name agent_system_name,
   in AgentSystemInfo agent_system_info)
  raises(EntryNotFound);

  Location lookup_place(in string place_name)
  raises(EntryNotFound);

  void unregister_agent(in Name agent_name)
  raises(EntryNotFound);

  void unregister_agent_system(in Name
  agent_system_name)
   raises(EntryNotFound);

  void unregister_place(in string place_name)
  raises(EntryNotFound);
};
```

Hereafter all methods of the interface are listed:

- *lookup_agent*(): this method returns the locations of specific agents; it can carry out the search by name or profile.
- *lookup_agent_system*(): this method returns the location of an agent system registered with MAFFinder; the search can be done by name or through the parameter AgentSystemInfo.
- *lookup_place*(): this method returns the location of a place registered in MAFFinder.
- *register_agent*(): this method adds a specific agent to the list of agents registered in MAFFinder; since an agent can migrate, this method must be frequently invoked in order to update the list.
- *register_agent_system*(): this method adds the specified agent system to the list of agent systems registered in MAFFinder.
- *register_place*(): this method adds the specified place to the list of places registered in MAFFinder.
- *unregister_agent*(): this method enables the removal of a specific agent from the list of agents registered in MAFFinder.

- *unregister_agent_system*(): this method enables the removal of a specific agent system from the list of agent systems registered in MAFFinder.
- *unregister_place*(): this method enables the removal of a specific place from the list of places registered in MAFFinder.

3.2 A possible implementation of MASIF

Let us suppose that Joe, a well-known manager, runs a chain of shopping centres. Joe, who in this example represents the authority, makes use of a mobile agents system to search for stock of goods in his stores (fig. 9).

Every store is represented by an agent system that keeps information about present goods in a database structure. Every time Joe needs to get supplies of goods, he connects from his client to MAFFinder that indicates him the nearest agent system from which he can create his own search mobile agent. Once the reference to an agent system is created, an agent is created in its turn; it, always through MAFFinder, migrates from one agent system to another looking for the information needed. Joe, besides, can constantly keep the agent he had launched under control, since the latter, every time it migrates from one agent system to another, leaves on the MAFFinder an address indicating the path to find it. The agent, once finished its task, can either go back to the client thus giving the search result, or stay "parked" in an agent system waiting to be resumed and called back by its own "creator".

4 FIPA

FIPA specifications are openly distributed to all society members. Since it is a technology that will further and unceasingly be developed, every distributed

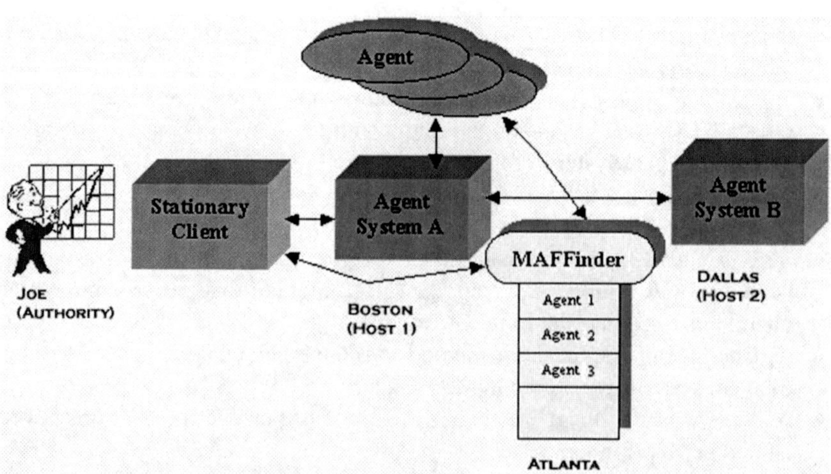

Figure 9: A possible implementation of MASIF.

specification is marked with one of the following states: Preliminary, Experi mental, Standard, Disapproved and Obsolete. Thus, it is possible to evaluate the reliability of specifications evolving in time and their advancement state.

Such specifications, however, do not entirely describe either the agent's internal architecture, or how it should be implemented; rather they try to explain which are the necessary interfaces to support interoperability among the various agent systems. FIPA architecture is based essentially on three points:

- specifications defining the architecture of the elements composing the platform and their relations (*Abstract Architecture Specification*);
- a handbook containing the specifications for the creation of particular agent systems and their communication technologies (*Guidelines for Instantiation*);
- the specifications defining agents and agent platforms interoperability (*Interoperability Guidelines*).

4.1 FIPA architecture

The architecture of platforms supporting FIPA communication protocol is defined by the *agent management reference model* (fig. 10). It gives models for the creation, record, search, communication, migration and elimination of mobile agents, i.e. all the series of operations enabling interoperability among different platforms.

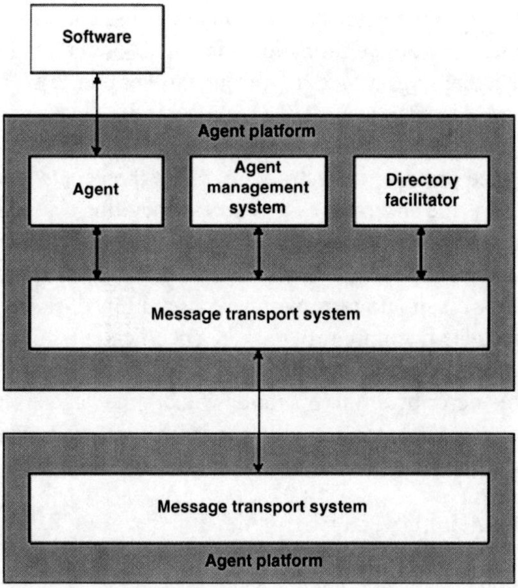

Figure 10: The agent management reference model.

Once this single background has been created, agents and platforms implementation can be achieved in various ways, using different programming languages. What makes the dialogue possible is the use of the same primitives for communication.

The agent management reference model consists of the following logical components:

- the agent is the main actor in an *agent platform* (AP), since it combines the potentialities offered by the service in a single execution model that, among other things, can interface itself with an external software and human users.
- every agent must have an owner at least (*authority in MASIF*), that could be a single person or a society for whom he works. Agent identification is also based on the knowledge of such information. The AID (*Agent IDentifier*) which it is associated to has the task of unambiguously marking it, so as to avoid conflicts in the use of names and in those operations, such as search, making a strict use of it.
- the *directory facilitator* (DF) is a special kind of agent providing a yellow pages service (i.e. it is intended for agents search). Thus, every agent can log its "position" by using DF, or invoke it a request such as to have information about the services offered by other agents. Inside AP multiple DF services can exist and work jointly.
- the *agent management system* (AMS) provides a service of agent names and also keeps an index of all the agents registered on an AP. Such index includes an unambiguous name for each agent; AMS is therefore a particular (kind of) agent with a control and supervision function on the access and use of an AP, such as an agent creation, elimination and record on the platform as well as the control of an agent migration to and from another platform.
- the *message transport service* (MTS) is the communication model used by the platform to make agents communicate.
- the AP gives the physical infrastructure where the agents can be executed. An example of AP is the ensemble of machines, operating systems or software supports able to receive mobile agents. However, we need not associate AP with a single host computer, since FIPA also supports the creation of platforms on distributed systems. It should be remembered that FIPA does not limit the creation of any AP, because it simply provides the specifications for agent-to-agent and agent-to-platform communication.
- the *software* describes all operations that the agent can execute; for example, the agent can interface with it to acquire new communication protocols, new safety algorithms, as well as access to packages supporting migration, etc.

4.2 Communication between two agents

Communication between two agents occurs through the exchange of messages encoded in a language called ACL (*agent communication language*).

FIPA ACL details communication between agents; a formal semantics composed of five levels is associated:

- *Protocol*: it defines the rules for the creation of dialogue among agents.
- *Communicative act*: it defines the kind of communication executed.
- *Messaging*: it defines some pseudo-information about messages including the sender and receiver agents' identity.
- *Content language*: it defines grammar and semantics to express a message content.
- *Ontology*: it defines the vocabulary and meaning of terms and concepts used in an expression.

FIPA specifications describe two services to achieve agents communication:

- a direction service (*directory-service*) enabling to carry out an agent search from its service description or its name;
- a service for message transport (*message transport service*) that carries messages from a platform to another.

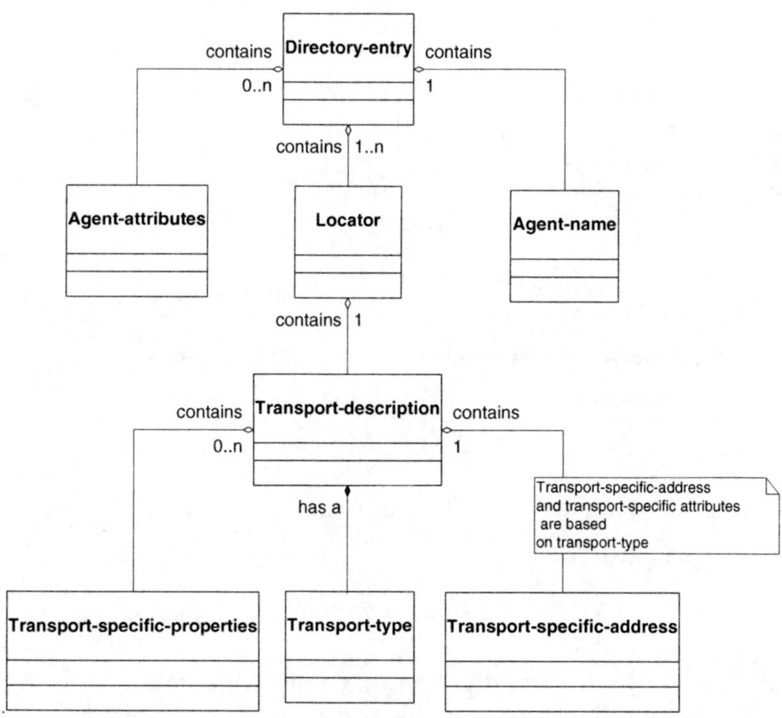

Figure 11: Information contained in the directory-entry.

4.2.1 Directory-service

The directory-service basic role is to provide the location in which to register the agents passage, so as to trace them back when searched for. Every agent registers itself in an entry list (*directory-entry,* fig. 11) as soon as it arrives; so in order to find an agent a directory-service can be used to go back to the directory-entry where it was registered (fig. 12).

Registration in the directory-entry is done by leaving a message in the form of a tuple composed of two basic fields, *Agent-Name* and *Locator* (table 3), plus two other possible fields describing, for example, the services offered by the agent, the costs associated to its use, etc.

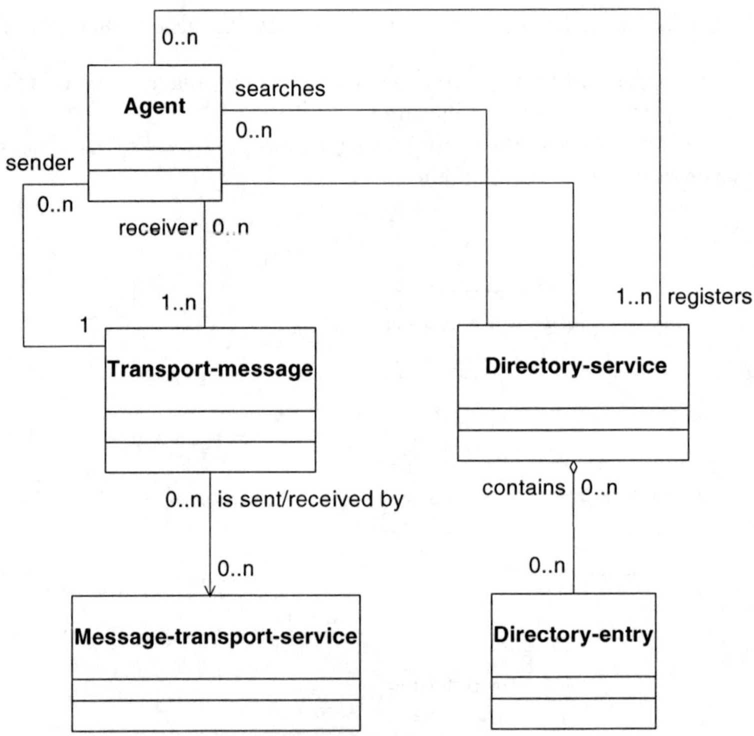

Figure 12: Relation among the main elements of FIPA architecture.

Table 3: The tuple structure in the directory-entries.

Agent-name	A single agent identifier (the way identifiers are created is described in FIPA specifications).
Locator	The description of how the agent's transfer must occur and communication with other agents.

4.2.2 Agent registration and search

Let us suppose that an agent A wants to make known its services and how to use them. First, it makes itself available for transport and reception, which can be achieved in various ways, according to its implementation; it could, for example, connect to an ORB, thus becoming a CORBA object, or connect to any list of agents available in the network.

Once the above-mentioned connection has been carried out, the agent makes its presence known by initializing a directory-entry and by registering itself there through a directory-service. Registration occurs by leaving the tuple containing information about the agent and the service it offers.

Now, any other agent B can use the directory-service to locate the agent A and communicate with it. The search can occur thanks to the service offered by the agent that is being looked for, or, more unambiguously, by using the identifier agent-name.

In the first case, the search result could be one or more agents all offering the same service. Therefore, there is the need to choose which agent it is most convenient to contact, on the basis, for example, of the cost of the service.

Fig. 13 and table 4 show the registration of an agent A and the search of it by another agent B, by using the directory-service.

Messages are implemented as tuples containing various fields (fig. 14), among which the sender agent's name (*Sender*) and the receiver agent's name (*Receiver*). Before being sent, they are encoded in ACL. Language content is described in a content language, such as KIF or SL, connected to a particular ontology containing all those objects that can be discussed in the message. Of course the names used for agents must identify them unambiguously, i.e. they must be generated using the functions defined in FIPA specifications for the assignment of a name to an agent. It is possible that a message contains another message.

4.2.3 Message structure

As we have already seen, communication in FIPA exclusively occurs through message exchange; it is therefore important to analyse the way such messages are created and finally sent.

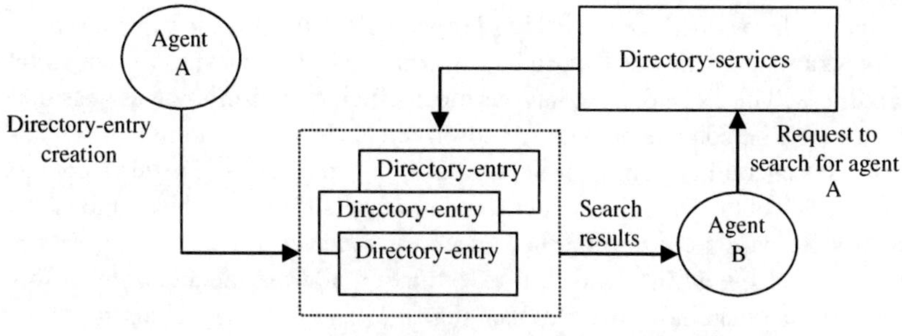

Figure 13: Registration in a directory-entry and agent search.

Table 4: Example of agent registration in a directory-entry.

Directory-entry for Agent1		
Agent name: Agent1		
Locator:		
Transport-type	*Transport-specific-address*	*Transport-specific-property*
HTTP	http:\\www.agents.net\agent1	(none)
SMTP	Agent1@agents.net	(none)
Agent-attributes: Attrib-1: yes; Attrib-2: yellow; Language: English, Italian, French; Preferred negotiation: contract-net		

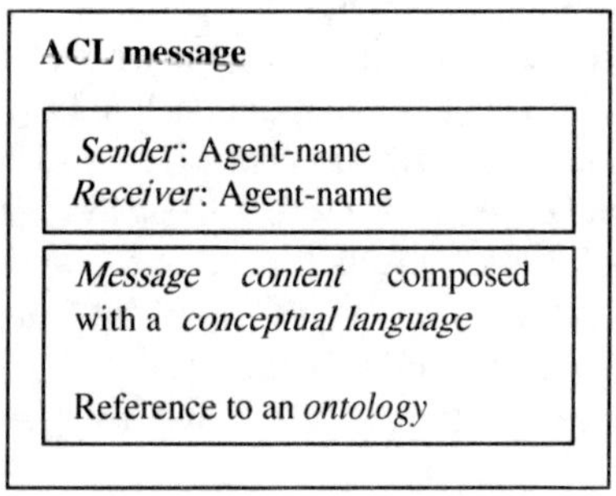

Figure 14: Message structure.

4.2.4 Message transport

Before being sent, the message is transformed in a *payload*, concerning the part of the message containing the communication content, and entered in the transport channel.

Finally, the payload is encoded in a proper code depending on transport modes. If, for example, a message must be sent on a wireless means, then a suitable encoding will be carried out to have as much efficient transmission as possible (in which case a bit conversion will be chosen rather than a string one).

Besides encoding, transport-message also puts the message inside a container or *envelope* containing both sender's and receiver's transport-descriptions. They are all information concerning the sender and receiver agents' migration modes (on which means do they travel, the destination address, details on how to use transport modes), how to send the message using the right communication protocols, and a series of additional information about, for example, how encoding was carried out, the security systems used, and other necessary information for transport (fig. 15).

As soon as the payload is kept in the envelope, message transfer can be carried out.

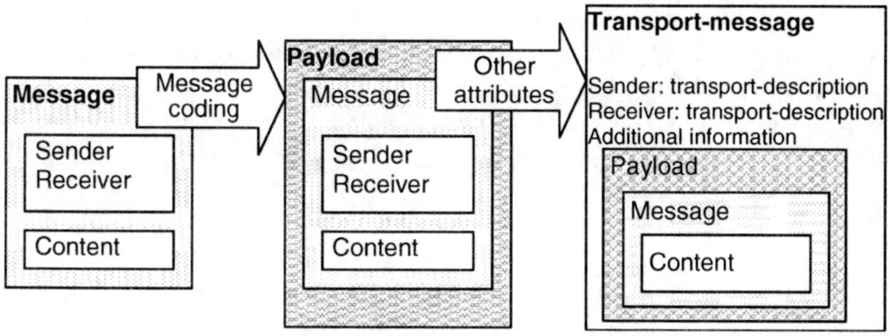

Figure 15: Message transport process.

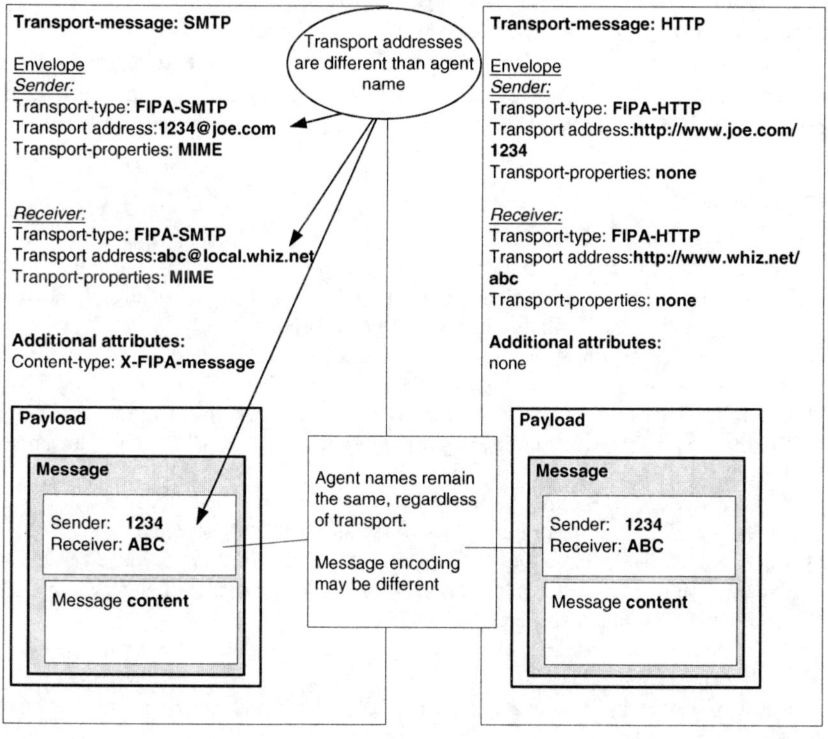

Figure 16: Difference in two message transports by using HTTP and SMTP.

4.2.5 How agents send messages?

FIPA agent platforms take into account agent–agent communication. It is achieved through the following basic notions:

- every agent has its own identifier AID.
- every agent can have one or more transport-description describing how to send or receive a message.
- every transport-description is related to a particular kind of message transport, such as IIOP, SMTP and HTTP protocols.
- transport-message is a message sent by an agent to another, encoded according to the type of transport that is being used.

Information concerning the transport-description ensemble can be kept in the locator.

Let us suppose, for example, that an agent has been identified with the name “Agent 1” and that it supports two message transport methods, HTTP and an SMTP mail address (fig. 16). “Agent 1” has therefore two transport-descriptions whose information are contained in its locator. Let us suppose the same situation as the one shown in table 4. Any other agent can now communicate with “Agent 1” by using one of the two transport methods described for “Agent 1”. Of course that implies the knowledge of all the characteristics of the agent it is communicating with, and therefore the possibility of suspending communication, to recover it later (carrying out, if necessary, a further search), or even changing communication method without losing contact, since the agent-name is always the same.

References

[1] Sumi, Y. & Mase, K., ART Media Integration & Communications Research Laboratories, IEEE AgentSalon: Facilitating Face-to-Face Knowledge Exchange through Conversations Among Personal Agents, 2000.

[2] Bellavista, P., Corradi, A. & Stefanelli, C., IEEE CORBA Solutions for Interopera bility in Mobile Agent Enviroments, 1999.

[3] Crystaliz, General Magic, GMD Fokus, IBM. Mobile Agent System Interoperability Facility, Available at ftp://ftp.omg.org/pub/docs/orbos/97-10-05.pdf, November 1997.

[4] GMD Fokus, IBM. Mobile Agent System Interoperability Facilities Specification, available at ftp://ftp.omg.org/pub/docs/orbos/98-03-09.pdf, September 1998.

[5] GMD Fokus, IBM, Supported by Crystaliz, Inc., General Magic, Inc. The Openg Group, Mobile Agent System Interoperability Facilities Specification, November 10, 1997. OMG TC Document orbos/97-10-05.

[6] CORBAservices: Common Object Services Specification, Revised edition, OMG TC Document 95-3-31, 1995.

[7] The Common Object Request Broker: Architecture and Specification, Revision 2.2, February 1998.

[8] Zahavi, R. & Mowbray, T.J., The Essential CORBA: System Integration Using Distributed Objects. John Wiley & Sons, Inc., 1995.

[9] OMG IDL to Java Language Mapping Specification, formal, 99-07-53, available at http://java.sun.com/j2se/1.3/docs/guide/idl/mapping/idltojavamapping.pdf, 1999.
[10] Kiniry, J. & Zimmerman, D., California Institute of Technology, IEEE A Hands-on Look at Java Mobile Agents, July–August 1997.
[11] IDL Type Extensione RPF, March 1995. OMG TC Document 95-1-35.
[12] Sun Microsystem, IDL Documentation, available at http://java.sun.com/j2se/1.3/docs/guide/idl/index.html.

Fault tolerance

Salvatore Geraci, Luca Giacalone, Carlo Leone,
Salvatore Mangano, Giuseppe Pitarresi, Alessandro Scaglione,
Salvatore Sorce and Alessandro Genco
DINFO – Dipartimento di Ingegneria Informatica

1 Introduction

The *fault tolerance* problem in mobile agent systems is a basic theme for any further development of those systems; a correct management of faults ensures in fact the agent's achievement of its tasks avoiding partial or complete loss of data and/or code.

Fault tolerance thus must necessarily be taken into consideration and dealt carefully since the probability of bad functioning increases with the complexity of operations, with execution time and number of processors involved in computation; faults can moreover occur both in the hardware and in the software, according to different modes and with various kinds of consequences.

In an agent system, where software entities act autonomously on behalf of a user and travel on a network of heterogeneous machines, faults can lead to partial or total loss of gathered data and/or of the agents themselves.

The ability to quickly detect, identify and possibly recover a fault increases the system's regularity and, therefore, its efficiency and reliability.

The first problem to deal with is to understand if the anomaly is really a fault or not. In agent systems on the Internet, where there are no limits to delays in communications and there is no relation among processors speed, it is difficult to determine exactly an agent loss: the person or application that created the agent can in fact think of an agent to be lost while it is late because of slow connection or processor [1].

Another problem is the fact that a network resource fault generates various alarms; that can cause a chain of faults ("domino effect") since the users sharing that resource can be many.

In order to deal successfully with such problems, various "fault management" techniques, both for hardware and software faults, have been developed, such as encoding techniques [2] or network configuration information [3]. The more popular SNMP [4] and RMON technologies [5], characterized by the centralization of fault management, do not enable to satisfy the exigency of scalability that today is a major one if we consider networks complexity.

2 Models of malfunction

Models of malfunction provide a way to discuss the behaviour of the system's components when malfunction occurs and to specify some assumptions about their effects. Such models enable to arrange malfunction from the weakest to the strongest one. According to such range some examples can be described:

- *byzantine* or arbitrary: components can have an arbitrary bad functioning;
- *timing*: it is assumed that a component will react correctly to an input, but not necessarily in the expected time;
- *omission*: a component erroneously reacts to an input;
- *crash*: a component has a malfunction that is not detected by the other components;
- *fail-stop*: similar to the crash, but detectable by the other components.

A system having a given model of malfunction can be represented as formed by components having a weaker model of malfunction.

3 Fault tolerant services

Some basic services offered by a standard hardware or by an operative system can be restructured so that they can continue to correctly operate despite the presence of malfunctions. Important services include stable storages, atomic actions and resilient processes (i.e. able to recover after a certain number of bad functioning).

The content of a stable storage is kept despite any bad functioning. Atomic actions enable to render a certain number of computations as an indivisible unit, independently from any simultaneousness or malfunction. A resilient process can be restarted and correctly continue the execution if the processor where it was being executed is bad functioning; some techniques to create resilient processes are for example checkpoints.

Another category of services for fault tolerance provides consistent information to all processes involved in parallel programs. Usually one wants to preserve the causality relation among the events occurring in the various processors in a distributed system. In order to do that, services such as a common global time, multicast information and membership are provided.

Common global time provides a distributed clock service that takes into account events causality even in the presence of bad functioning. A multicast service sends messages to every process in a group of processes according to a prearranged order, independently from any bad functioning or from the competition among the processes themselves. Finally, a membership service provides the processes belonging to a group with consistent information about the ensemble of functioning processors at a given time.

4 Structural principles of programming

Fault tolerance structural paradigms are program structuring canonical techniques that have been developed together with the previous services and abstractions in order to help the programmer in structuring fault tolerant distributed programs. There are three program-structuring paradigms for fault tolerant softwares:

- object/action;
- primary/backup or restarting action;
- replied state machine.

In the first paradigm an application consists of objects and actions. An object encapsulates critical data in a local state and exports certain operations to modify data. It is assumed that data have a long life and are recorded in a stable storage. Actions are operations modifying the object state to which they belong. Their execution is transactional, i.e. those actions can be serialized and retrieved notwithstanding malfunctions. In an action that can be serialized the effect of any concurrent execution of actions on the same object is equivalent to some serial sequence.

In the primary/backup paradigm, the execution system periodically saves processes local state into the permanent storage. Checkpointing and rollback scheme is the most widely used technique for the implementation of such paradigm. Primary process is active, while backup processes are passive. Only active process responds to the service demands; when primary process does not function properly, one of the backup processes will turn a primary one, starting from a state saved in the last checkpoint.

In the replied state machine paradigm, an application is structured as a set of services, each of which is implemented as more identical deterministic processes. Each service demand is sent to all the processes providing the service. Each process operates as a state machine modifying its state variables as an answer to the commands received by the other state machines or by the environment. So, if the execution is correct, every process has the same state. If some states do not match, it is considered valid the one that is common to most processes.

Since every service processes each command, the state machine approach is sometimes called active replication, while the primary/backup paradigm is called passive replication. Both approaches need nevertheless redundant processors.

5 Languages for fault tolerant programming

Several fault tolerant programming languages have been developed in order to help creating fault tolerant programs. Some of them are Argus, Avalon, Fault tolerant Concurrent C, FT-Linda, Orca, FT-SR and PLinda.

Generally speaking, those fault tolerant programming languages differ because of the supported programming structural paradigm. Argus and Avalon support the object/action model; Fault tolerant Concurrent C and FT-Linda support the replied state machine paradigm. FT-Linda is a fault tolerant version of Linda for parallel applications. Instead of using state machine paradigm to replicate processes, FT-Linda uses it to replicate Linda's shared memory.

Programming languages supporting restartable action paradigm often make use of mechanisms for process checkpoint on disk, and rescue action is based on the most recent checkpoint. Language execution system can manage checkpoint and rollback in a transparent way to the programmer. Orca is a language that autonomously performs parallel applications checkpoint. It uses a broadcast in order to ensure that checkpoints global consistency is maintained.

FT-SR is a language projected to support several programming paradigms. PLinda is a set of Linda extensions projected to support strong parallel calculation and the use of idle workstations.

6 Fault tolerance through mobile agents

In client–server applications, servers typically provide a public interface through a prearranged set of primitives. Clients can call for high-level functionalities composed of such primitives, and their needs can change in time. Rather than modifying the server's interface to support each client's exigencies, a client can keep its interface to the server's node, by using a mobile agent. This feature moreover curtails the number of network-based interactions. The service providers can take advantage of that to dynamically improve the server's abilities.

Mobile agents are intrinsically stronger than remote procedure call systems, since they do not depend on network availability, and when the agent migrates to a server machine it is not influenced by the client or network temporary bad functioning.

Since mobile agents are concurrently executed, they are also useful because they can provide a mechanism for the introduction of parallel activities. A client can split its tasks through multiple agents in order to provide the application with parallelism or fault tolerance.

Most publications on this topic refer to particular agent systems, or deal only with some aspects concerning mobile agent.

Among them Johansen [6] developed a system of fault tolerance management concerning agent migration in TACOMA agent system [7]. When an agent migrates, a *rear guard* agent is created in the source node monitoring the agent's migration. This very simple concept does not take into consideration network segmentation management.

As far as the TACOMA project is concerned, Minsky [8] as well proposes an approach based on the concept of "itinerary", for example an *a priori* planning of the nodes the agent must visit and the work it has to do there. In that model the itinerary is assumed to be already known and the order of nodes to visit prearranged. Fault tolerance is obtained by simultaneously executing every step of the itinerary on multiple nodes and by sending results to every nodes of the following step. The disadvantages of this fault tolerance model are the itinerary high rigidity and the extreme simplicity of agents. For example communication among agents – indispensable in the case of parallel and distributed applications – is not taken into consideration.

Strasser and Rothermel introduce a more flexible itinerary approach [9] in Mole environment [10]. The whole agents work is divided into stages, each of which includes the action that must be performed in a node. When a mobile agent enters a new stage, migrating to the following itinerary node, called *worker node*, it is also duplicated on a number of additional nodes, called *observers*. If the worker turns non-valid (due to network or the same node fall), the highest priority observer is selected as a new worker thanks to a particular *selection protocol*. A *voting* protocol ensures the stopping of several damaged workers in case of network fault. The last protocol is integrated with *2-phase-commit* (2PC) protocol. It causes though excessive network overhead. As far as communication among different agents is concerned, nothing is stated in detail.

Vogler [11] focuses his work on the realization of an algorithm for a more reliable migration based on the already mentioned 2PC protocol. He deals with no other aspects apart from this one.

Murphy and Picco [12], on the other hand, deal with only a particular aspect of agent systems: the problem of messages never reaching their destination because of the agent's high mobility; problems concerning undelivered messages due to network fault are not treated.

Mobile agent technology finds nevertheless applications in the management of heterogeneous distributed clusters (FLASH), in the paralleling of calculations (mobile agent team system, MATS), in fault tolerant parallel-distributed calculation (James and A^3).

Before investigating some fault management algorithms thoroughly, we will discuss various kinds of faults occurring in mobile agent systems (such as the non-delivery of a message, possible transmission faults, impossibility of executing code), mentioning their possible consequences.

7 Possible faults

Current operating systems do not take into account the use of mobile agents; therefore, there is the need of dedicated systems acting as brokers between the operating system and the agents. These platforms functions are widely discussed in papers, for example in Ref. [13]. Although the problem of fault tolerance is dealt with in various research fields, the variety of faults occurring in mobile agent environments is such that it is difficult to find solutions managing them all; rather, the various suggestions focus only on some aspects, neglecting others.

7.1 Fault of a node (site)

The complete fault of a node implies not only the loss of all its data (files, databases, the operating system itself and possible applications useful to agents), but also the unavailability of that node's agents. Depending on the fault being temporary or permanent, various issues will be dealt with and managing modes will be therefore differentiated. Of course, in the event of a permanent fault the main problem would be that of complete impossibility of retrieving the data gathered by the agent. Solutions for this case are centred on the precautionary and temporary saving of data. Suggestions concerning this are many: from the use of checkpoint to that of agent's clones, to centralized techniques based on a manager agent, to the saving of all data in a dedicated server or even in all the sites visited by the agent (in which case we talk about the agent's "traces"). This kind of fault management has also the need of implementing agents not bound to a single itinerary. When the fault is temporary, the fault manager will deal with the restoring of data or with their deletion, if they are thought of as redundant and unnecessary. If the creation of clone agents is expected, it is necessary an algorithm for the search of "twin" agents, as well as one for their data comparison choosing, according to pre-established criteria, which ones to delete or not.

7.2 Fault of an agent system components

In such cases sites cannot fault, but parts of them or, more precisely, parts of the system offering services and functionalities for the achievement of the agents' tasks. As a consequence, the agent cannot work properly and, therefore, data (and results) inconsistency or a reduction of the agent's functionality occurs. The management of those faults needs that each site that can be visited has also the platform containing all the functionalities needed by the agent. Thus, it is always possible for the other nodes to find unavailable services.

7.3 Agent damage

Besides suffering from the total or partial fault of sites, the agent's behaviour can be "defective" both for its own nature (wrong implementation) and for problems of different nature during execution on a site. The agent, for example could need more resources than the ones actually available or cause a deadlock or even be deleted. While deadlock situations should be managed by the operating system (local solutions therefore), as to the request of more resources there exist publications with resource balancing algorithms and related load balancing.

7.4 Network breakdown

The entire communication network or a single link breakdown can lead to a node's isolation or to the network segmentation. This kind of faults happens very rarely. It is, therefore, costly to implement a fault management algorithm in this case, if we further consider that, from the agent's point of view, this is a case equivalent to the one of a site's fall.

7.5 Message falsification or loss

These faults are usually caused by defects in the network or in the communication unit of the agent system. The agents' impossibility of migration can also fall within this fault category. These faults are very important when agents have the necessity of communicating and cooperating among them or with remote sites, in order to carry out their task. A wide range of articles and papers can be found in available literature (e.g. in Ref. [12]).

8 Conditions and requisites for a fault tolerant execution

In order to ensure a correct approach to fault tolerance, mobile agent systems and agents themselves must have some characteristics:

- *agent communication support*: it is essential for distributed parallel applications;
- *autonomy*: if the actions to ensure fault tolerance were taken on by a supervisor, autonomy would be limited. Each decision taken by the agent according to its natural autonomous behaviour should in fact be coordinated with the supervisor. Apart from fault tolerance, other aspects would be involved, since possible actions could contradict the supervisor's decisions;
- *fault tolerance as optional characteristic*: not every mobile agent environment application need fault tolerant execution. The user, the agent environment or the applications themselves should be able to individually decide whether and when fault tolerance should be activated. Fault tolerance activation for an application, moreover, should not affect another one already in execution, so that the possibility of simultaneously executing applications that require it and others that do not require it can be ensured;
- *transparency to the user*: application programmers and users are usually interested in encoding their algorithms and solving their problems. They would not like to deal with fault tolerance in their applications. Therefore, its execution should not modify their applications encoding;
- *efficiency*: one of the main goals of using mobile agent techniques in distributed programming is an efficient use of resources. Therefore, fault tolerance must not drastically increase overhead in case of operations lacking/without faults. That is a further reason why fault tolerance should be an optional characteristic. Moreover, recover in case of fault should be achieved as quickly as possible;
- *hardware, operating system or "run-time" environment*: the use of fault tolerance method should not require changes to the hardware, the operating systems and the dedicated environments; besides, portability must be guaranteed;
- *portability and re-usability*: the changes in an agent environment should imply additions or adjustments, not re-planning or re-implementation. Existing functional units should remain usable at least for executions without fault tolerance. Such units' re-use ensures further execution of pre-existing applications.

9 Fault tolerant mobile agent

To make an agent system have the above-mentioned characteristics, the system and the agents themselves must satisfy the following requisites:

- *modular architecture*: by using a monolithic agent, it could be impossible to adopt specific fault tolerance functionalities. If fault tolerance request occurred during run-time, it would be necessary to replace the agent by one carrying out the same tasks, but possessing fault tolerance functionalities. All that would lead to the use of further memory (useless if no faults occur). In order to avoid that, it is common to organize the agent's structure in functional modules;
- *separation between applications and agent's kernel*: the difference between applications and agent's kernel on the one hand facilitates transparency to the user; on the other hand, it gives the agent the opportunity to execute different application modules during its life-cycle. Applications, though, must have the possibility of modifying the agent's behaviour;
- *parallel functioning of functional modules*: such an organized structure suggests that functional units should be contained in modules working in parallel. This makes the application of different strategies independent and concurrent;
- *adaptability*: there ought to be the possibility of influencing the agent's behaviour during its execution. Without this feature, it could be impossible to activate the fault tolerance required by the user;
- *automatic survey of dependences*: we have already seen that the agent must be structured in a modular way. That ensures that only the modules necessary to carrying out the task can load the agent. When a module is replaced, it can happen that the new one needs some service by another module not present in the agent yet. In order to maintain transparency to the user, such dependence must be automatically satisfied.

Let us suppose we are in the presence of an asynchronous distributed system, without limits in transmission delays and with different speed processors, such as the Internet.

A mobile agent will function on a sequence of machines, where a position P_i $(0 \leq i \leq n)$, also called *place,* will provide for the execution of the agent logical environment. We will call the places where the agent's first and last stage occur *source* (*agent source*) and *destination* (*agent destination*), respectively.

The sequence of places between source and destination is called itinerary of a mobile agent. While a static itinerary is defined in the source and cannot be modified by the agent's execution, a dynamic itinerary is subject to changes by the agent itself. From a logical point of view, a mobile agent carries out its work in a sequence of "stage actions" (fig. 1). Every stage action potentially consists of multiple operations. The agent (a_i) to the corresponding stage S_i, represents the agent that has executed the stage actions on places P_j $(j < i)$ and is in execution on place P_i. The execution of (a_i) on position P_i will lead to a new inner state of the agent, and it can as well happen that the position will find itself in another state

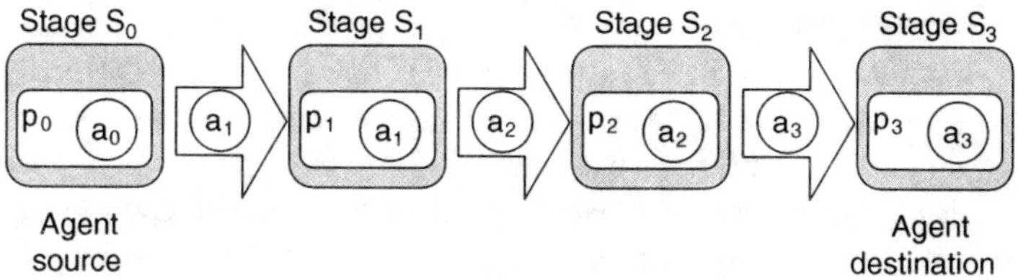

Figure 1: Sequence of stage action.

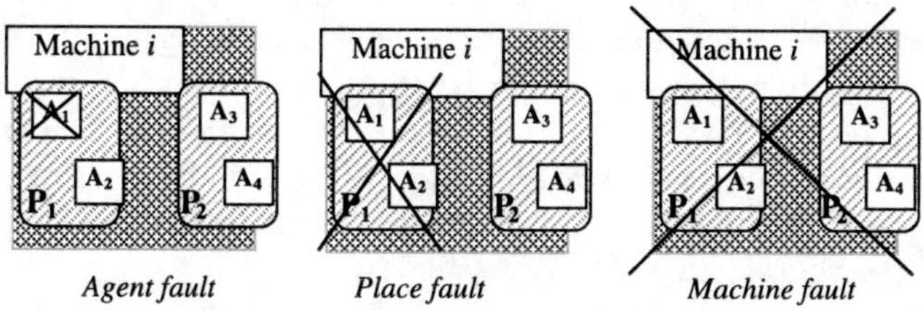

Figure 2: Kinds of fault.

(e.g. if the agent has had effects on it). The resulting agent will be indicated with (a_{i+1}). Place P_i will forward A_{i+1} in P_{i+1} $(i < n)$.

Machines, places or agents can fail by crashing. In this section, we are going to examine crash failures, while other artificial damages, caused, for example by programming errors or violation of security systems, are not taken into consideration. A fault in the place causes the fault of all its agents. Likewise, a fault in the machine causes the fall of places as well as that of their agents. Last, a fault of connection causes the loss of messages or agents passing through it in that moment (fig. 2).

Faults in an environment where mobile agents are being executed cause the partial or total loss of the agent; that implies the loss of the agent's code and state. Although the agent's owner can keep a copy of the code, when fault occurs the agent's intermediate state would nonetheless be lost and, together with it, the already executed stage actions results. But, even worse, the agent loss can lead to some inconsistencies in the whole system. Let us assume, for example an agent that recovers money from the agent's owner's bank account: the agent loss deletes every reference to the recovered amount of money.

In an asynchronous system, the agent's owner cannot distinguish between the agent loss and its delay due to slow connection or processor [1].

One of the following cases can occur:

- the agent got lost whereas the owner thinks it is just slow: the owner waits in vain for the agent's coming back;
- the agent is late whereas the owner thinks it got lost: the owner sends another agent that provokes its duplication and, therefore, a double execution.

Both cases have an unwanted effect on the mobile agent execution: in the first case the owner keeps doing his activities unaware of what happened; in the second case, on the contrary, he provokes the agent's code multiple execution.

Let us assume, for example that the agent's task is to buy a plane ticket. The agent's code multiple executions can lead to the unwelcome purchase of several tickets.

Generally speaking, stage actions having side effects (i.e. modifying the state of the place where they have been executed) must be executed just once [14]. Otherwise, operations such as bank account readings can be executed several times without changing the application semantics (a reading operation as a matter of fact does not modify a place's state).

The mobile agent's fault tolerant execution prevents the agent loss, ensuring it reaches its destination; thus, doing the owner's uncertainty is eliminated.

Thus, the agent's arrival to its destination is ensured and case 1 cannot occur. Moreover, it is not necessary a new sending of the mobile agent, as in case 2.

Even though the agent loss has been prevented, the case of a fault in a place must as well be considered.

The term *blocking*, when a fault occurs in the link or the place, indicates the case in which the agent's execution progress is hampered. Furthermore, since fault tolerance can only be carried out by having recourse to certain forms of redundancy, the risk is to have several instances of the agent executing simultaneously.

The properties required for the agent's fault tolerant execution are the following:

- *non-blocking*: the mobile agent must not block;
- *exactly-once*: stage actions must be executed just once.

The first property states that the agent's execution must not be interrupted. A blocking cannot be prevented if all places undergo a fault at the same moment. Every approach to fault tolerance has, therefore, higher limits as far as the number of tolerated fault is concerned.

The second property requires that the agent's operations be executed just once. It must obviously be noticed that agent's operations not always need to be concluded, for example in the case in which our agent must ask for unavailable plane tickets. Nonetheless, the agent's action (plane ticket demand) respects the exactly-once property.

10 Checkpointing

A periodical check of the agent's state and code on current place can prevent its loss. While (a_2) is in execution on P_2, its code is initially recorded in a safe store, while its state is periodically checked. After a fault of P_2, (a_2), it is recovered from the last checkpoint and its execution can go on. Nevertheless, while P_2 is still in fault, the agent's execution is blocked. Vogler [11] prefers the following approach. In order to prevent the agent loss during the transmission between P_1 and P_2, Vogler makes use of a communication mechanism based on negotiations between the two places.

11 Replication

Replication prevents both the agent loss and the problem of blocking. By adding some redundancies, we will hide faults and make it possible for the agent to carry on its execution notwithstanding faults. We will make a distinction between temporal-replication-based (TRB) and spatial-replication-based (SRB) approaches.

TRB approach is based on *checkpointing*. Instead of storing the agent's code and state (a_2) in the place P_2 (fig. 3), a copy of it is kept in the previous place P_1.

The storing of (a_2) in the previous place thus prevents the agent loss. It is furthermore possible to send (a_2) in the place P_2^I, if a fault has occurred in place P_2. That is the reason why P_1 checks its succeeding ones: places P_2 and P_2^I.

TRB approach thus avoids the blocking situation that occurred in the Checkpointing scheme.

In SRB approach, otherwise, a set of places $M_i = \{P_i^0, P_i^1, P_i^2,...\}$ execute the agent at S_i stage. Place P_i (that will have finished stage actions) sends simultaneously a replica of the agent in every place of the set M_{i+1} of the following stage S_{i+1} (fig. 4).

Then the places in M_{i+1} will execute the agent (a_{1+1}) and will carry it to the next stage.

Even if a place fails, the agent (a_{1+1}) is not lost, since the other places in M_{i+1} have also received (a_{1+1}). At the same time blocking is avoided, because another place acquires the execution of (a_{1+1}). Ideally, therefore, the mobile agent is executed only in a single place (logical). Execution in all places generates widely unwelcome considerable overload.

Schneider's approach [15] effectively executes the agent in all stage places (we will see later that is not necessary). Such a redundant execution of mobile agent is fundamental for invulnerability to Byzantine faults, for example errors related to safety, in particular against attacks from mobile agent's malicious hosts.

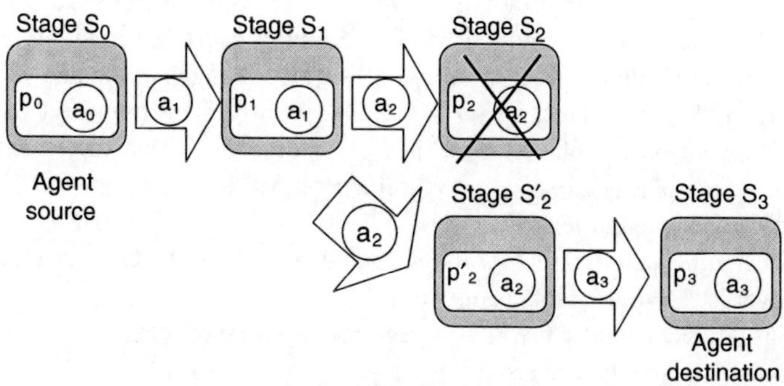

Figure 3: TRB approach.

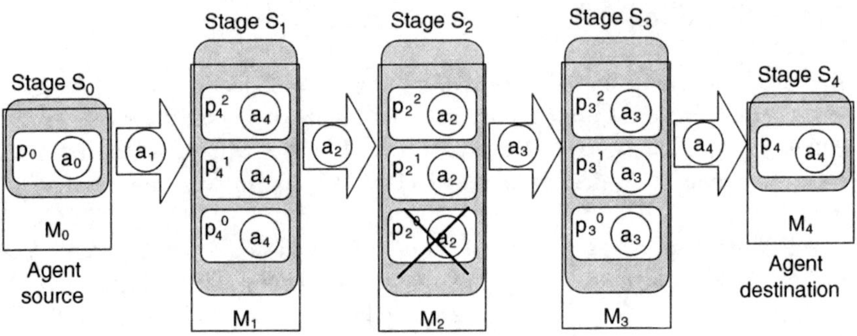

Figure 4: SRB approach.

11.1 Place replicas

Both TRB ($M_i = \{P_i, P_i^I, P_i^{II},...\}$) and SRB ($M_i = \{P_i^0, P_i^1, P_i^2,...\}$) approaches count on place redundancy. While in TRB approach such a redundancy avoids blocking, in SRB approach it serves both as a means against mobile agent loss and against blocking. In TRB approach, in fact, the agent loss is not avoided by redundancy, but by a copy of the agent stored in the previous place.

As far as relations among places are concerned, three different classes of place can be distinguished [9, 16]:

a. iso-place;
b. hetero-place;
c. hetero-place with witnesses.

The *hetero-place* class finds out a set M_i where all places provide a similar service such as, for example the sale of plane tickets from Palermo to Rome. Anyway, places are provided by different airlines, such as Alitalia, Meridiana, etc. *Iso-place* class specifies the traditional case of server replication.

Looking again at our previous example concerning airlines, set M_i is composed of replica places provided by the company itself, for example all places are served by Alitalia and are exactly replicated. Changes in a place will be visible also to all other places in set M_i. Consequently, fault tolerant execution of a mobile agent on iso-place leads, in SRB approach, to two replication levels: replication on place server (e.g. Alitalia's or Meridiana's servers) and client replication at the agent level.

In TRB approach two replication levels occur only in the case in which a fault is discovered. Iso-places are assumed in Ref. [15].

Let us consider finally the *hetero-place with witnesses* class, which is a generalization of hetero-place class.

While all hetero-places provide a particular service (e.g. Palermo–Rome plane tickets), in the *hetero-place* with witnesses class just one place subset provides such service. The other ones (the witnesses ones), even if they can execute the agent, will not provide a plane ticket, so the service required by the agent fails. However, the agent is not lost and goes on with the execution, while it will potentially report the failure in obtaining the ticket to the agent's owner.

The following developments of mobile agent fault tolerance are mentioned in order to complete this analysis.

They are generally applicable to all mobile agent fault tolerance systems.

Besides place redundancy, NetPebbles [17] proposes the so-called task-level-redundancy. A task matches a mobile agent stage action. Strasser [9] suggests extending a mobile agent fault tolerance by providing alternatives in the itinerary place order; more in detail, the lack of a fixed order in the itinerary enables the agent execution first on places without faults.

Anyway the agent needs execution on those places where fault occurred but, in the meanwhile, they could have been already repaired and available again. Such an extension is of course applicable only to itineraries where computation order makes no difference.

12 Exactly-once execution property violation

While replication prevents blocking problem, it does not ensure the exactly-once-execution property. Actually replication, both in TRB and SRB approaches can lead to multiple executions of the agent.

Let us consider, for instance, TRB approach in fig. 3. When P_2 is available again, its consistent state must be recovered. More precisely, all stage operations of A_2 executed on it must be cancelled; otherwise, we would have violated the exactly-once property, since (a_2) was executed on P^I_2.

Otherwise, the effects of a stage action of (a_2) can be seen both in P_2 and in P^I_2, thus, violating exactly-once property. The same applies to SRB approach.

Consequently, the agent's stage actions will need to run as *local transactions* (local to the place). Therefore, stage actions will only be "commissioned", making themselves visible (to other places) when a potential violation to such property is solved. Transactions, therefore, are operations that must operate on several resources on a distributed basis, maintaining global consistency. The properties a transaction must take into consideration are typically indicated by the acronym ACID:

Atomicity: all transaction operations are successfully executed; otherwise none must be executed;

Consistency: the set of executed operations must leave the state of the system in a consistent situation;

Isolation: the effect of transaction operations must not be visible to objects external to that transaction until it is not completed;

Durability: transaction effects must be durable and continue also after the end of it.

At local level already, the achievement of such properties implies synchronization problems that can be solved through the acquisition and release of resource locks, minding to avoid (optimistically or pessimistically) possible deadlock situations.

In a distributed area, therefore in our case, i.e. where transaction involves several processes even on different hosts, issues proliferate; in particular, we must take into consideration the case of communication errors and possible fall of a participant.

By using the technique of *local transactions*, the agent partial execution can be cancelled without influencing the other places and, therefore, it is possible to bring a place in state of fault back to a consistent state.

An interesting case occurs when an agent generates a child (spawn operation), i.e. when another agent starts (fig. 5). The cancellation of the agent's child-generating operation consists in the cancellation of it and of all the activities executed until then by it.

The cancellation of a child agent's stage actions, after he has already left current place, is a very complex operation. One could think of a technique of message exchange, such as undoing ones. They (the messages) are sent to the agent, but they could never reach it, since it is potentially able to move faster than the message itself. From now on we will assume that a generated child agent is immediately created but starts working when its parent agent's stage actions have been concluded (in which case we are sure that the child agent has been executed by an agent that has effectively concluded its stage, and therefore is not a partially executed agent, or a replica).

The second possible violation source of such property is a wrong fault determination.

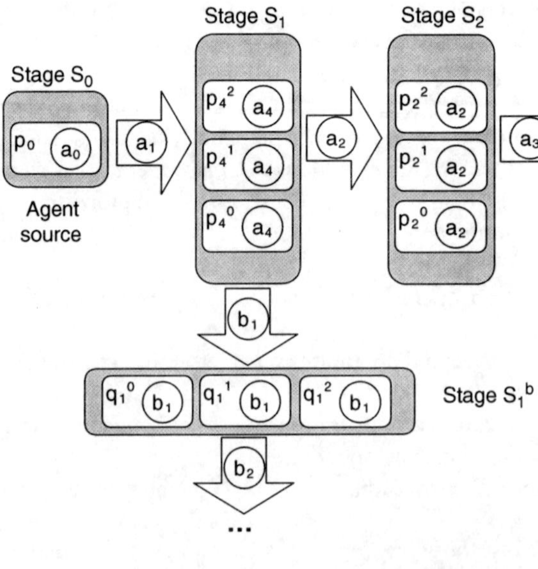

Figure 5: Child agent generation.

That generates misunderstandings between two places, i.e. it can occur the hypothesis that the other place seems to be faulty, but it is not and continues executing the agent.

13 TRB

As shown in fig. 3, place P_1 keeps a copy of A_2. When it detects a fault of the agent (a_2) on place P_2 it sends such copy to another place. The agent's current execution monitoring enables managing a certain number of faults (it depends on places availability) sequentially occurred in P_2, P^I_2, P^{II}_2,....

Let us assume that also P^I_2 has a fault. That fault will possibly be detected by P_1 and another copy of (a_2) will be therefore sent to the other place P^{II}_2. Despite that, a simultaneous fault of P_2 and P_1 will provoke the agent's loss and so its execution blocking.

NetPebbles [17] overcomes that problem by introducing a monitoring scheme where previous stage's places monitor the next places. Every place sends "heart-beat" messages to the previous place, within a certain distance. Such distance is defined as the difference between the indexes j and k of the two places P_k and P_j (with $j > k$). "Heart rate" decreases with the increase of distance, in other words the greater the difference is between indexes j and k, the lower the frequency that P_j uses to send its P_k its beats is.

Place P_k will send the agent (a_{k+1}) in another place P^I_{k+1} if and only if it suspects that all subsequent places are in fault, for example if it has not received heartbeat messages anymore. That enables NetPebbles to manage a simultaneous number of faults equal to the value of distance.

Through other means, James [18] uses a selection protocol to determine place P_l that keeps memory of the agent's more recent copy (for example (a_l)). Place P_l will then send a copy of the agent to place P^I_{l+1}, different from P_{l+1}, where the original agent had been sent. The agent's state stored copies, as well as the code of multiple executions; need a great amount of space. Whereas mobile agents have usually a small size, most of them can ask the places a considerable amount of space.

13.1 Exactly-once property in TRB

TRB approaches lead, potentially, to the duplication of agents. Let us assume that P_k erroneously suspects a fault of (a_{k+1}), shown in fig. 6 by a cross. The sending of (a_{k+1}) to another place P^I_{k+1} leads, therefore, to several instances of the agent: the former agent (a_{k+2}) and the duplicated one (a^I_{k+2}). Since a mobile agent itinerary can dynamically vary, the copy agents can follow completely different itineraries (a_{k+2}) than the former agent's ones.

Consequently, the copy agents (a_{k+2}), (a^I_{k+2}), ... will be discovered only when they reach the same place (usually the agent's final destination). At the same time, other agents will probably have had access to the results of the agent's execution on the places, and therefore they will read invalid data.

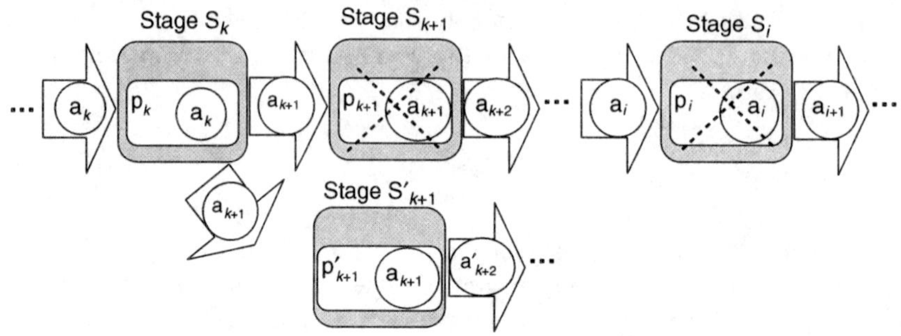

Figure 6: Exactly-once property in TRB approach.

TRB approaches, therefore, need to obtain the agent's executions isolation. By executing stage actions as local transactions, the agent prevents the access to all the data items used. Access to data tried by other agents will be blocked until the agent's current execution is over. The agent's instance stage actions will therefore be executed, while other instance's ones will be rejected. Thus, the agent's execution isolation is achieved.

Data items must necessarily be blocked until the end of the agent's execution, even if it is not in fault. Let us assume, for example that the agent is in execution on stage S_i: due to network partitions or slow connections, place P_k could receive no more heartbeat messages from other places P_j ($j > k$) and therefore it will suspect the fault of all its successors. It, therefore, will send a copy of the agent A_{k+1} to place $P^{!}_{k+1}$. Hence, there will be a copy agent $A^{!}_{k+2}$, though the original agent's execution had much earlier executed the stage S_k and is currently in execution on P_i (with $i > k+2$).

By blocking the data items used by the agent, the system's throughput will deeply be limited. Moreover, once the agent has reached its destination, it is required the sending of extra messages to all the itinerary places concerning the stage actions commit or abort. That problem has not been dealt with yet by Refs. [17] and [18].

In James [18], a fault tolerant search tool (lookup directory) prevents the agent duplications: it is a replicated device and enables the synchronized access to its methods. Before the execution of an agent (a_i), place P_i accesses such device and checks that (a_i) has not been executed on another place $P^{!}_i$ yet. When the execution of (a_i) has ended, P_i sets up the corresponding access to the search device; thus, indicating that the stage action of (a_i) has been concluded.

Let us assume that the agent is currently in execution on place P_{i+1}; from here the corresponding access to the search device indicates that the execution of A_i has been concluded. If P_i and P_{i+1} are suspected of faults by previous places, a selection protocol execution will identify P_{i-1} as the place with the most recent available agent's state (i.e. (a_{i-1})).

James does not ensure the mobile agent exactly-once execution; rather he makes use of *at-most-once* or *at-least-once* property semantics for stage actions. These semantics are weaker than exactly-once property.

An exactly-once stage action is equivalent to the stage action satisfying both semantics. Moreover, two other semantics are provided as far as the agent's execution itinerary is concerned: the *atomic* and *best-effort* ones, i.e. either the mobile agent will have to be executed on all the places of its itinerary, or on the majority of them.

14 SRB

The approach based on spatial replication overcomes the problem of blocking. In such approach, the set of places M_i of a stage S_i is responsible for the fault tolerant execution of the agent (fig. 4). Redundancy is not required to the agent source and destination. In order to mitigate the consequences of machine faults, if a machine runs stage S_i, every place P_1^j ($j = 0, 1, \ldots$) is usually put on different machines (even though that is not necessarily required). Works in Refs. [16, 19, 20, 21] defend the use of SRB approach for the fault tolerant execution of mobile agents.

Revisiting the former example of a fault in a place, fig. 4 shows that the agent's execution can now proceed to places P_2^1 and P_2^2, even if place P_2^0 is in fault. Since all M_2 places have received A_2, the agent will not be lost even in case of simultaneous fault in all places except one. Moreover, places redundancy ensures that the agent's execution is not blocked. Set M_i dimension defines the number of faults that can be simultaneously managed without risk of blocking.

14.1 Exactly-once property in SRB

If on the one hand, SRB approach ensures the agent to reach its destination and solves the problem of blocking, on the other hand, it does not prevent multiple executions of the agent. In fact, a wrong suspect of fault can lead to the agent's execution on more than one places of M_i. Let us assume, for example that P_i^0 is suspected by P_i^1 due to a rather slow connection or to network partitioning (fig. 7). Place P_i^1 will start the execution of (a_i), which will lead to (a_{i+1}) and M_{i+1}. In the meanwhile, the execution of (a_i) on P_i^0 will lead to (a_{i+1}^1) and M_{i+1}^1. Since M_{i+1} and M_{i+1}^1 are usually different sets, the duplicated execution of the agent cannot be detected but at the end.

In order to limit a duplicated agent lifetime, SRB approaches usually provide a mechanism apt to make their determination possible. It limits the duplicated agent's life, which will last at the most the time necessary for the execution of a stage. After the stage execution, its copies are discarded and their actions brought back to previous state, using local transitions. Current SRB approaches are very different as far as their way of determining a duplicated agent is concerned.

In order to preserve the exactly-once property, some SRB approaches select, among the set M_i places, a place called *worker* that is responsible for the agent's

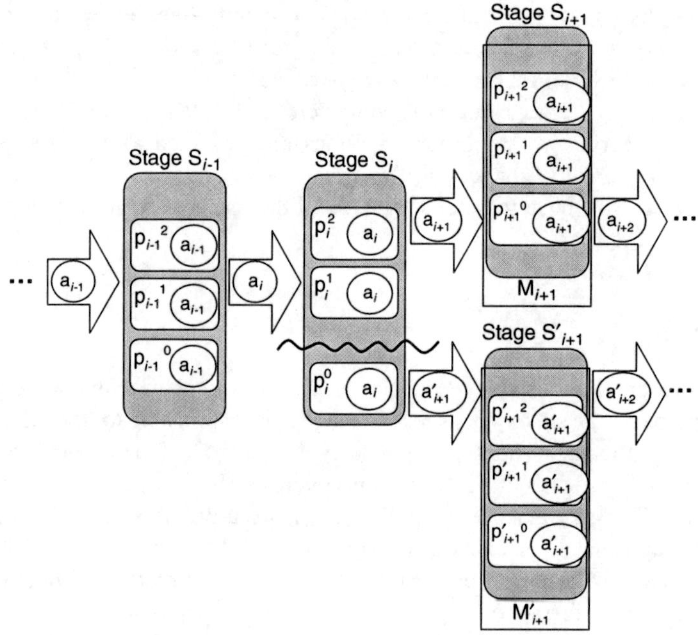

Figure 7: Network partitioning.

execution. The other replica places receive a copy of the agent, but function as the *worker* place observers. When it is in fault, *observer* places will determine that and will execute an election protocol to select a new worker place. If it is not able to provide the service required by the agent, it will execute a mechanism of exception managing.

The simplest approach is that of relying on a reliable fault determination system. In other words, if a place is suspected, it will effectively be in fault. Therefore, false fault determinations cannot occur.

Perfect fault determination is necessary but not enough to prevent the agents' duplication. Let us suppose that P_i^1 determines a fault of P_i^0. It will never know if P_i^0 has succeeded in sending the agent to the next stage or not. Consequently, NAP [21] makes use of a reliable broadcast technique ensuring that all places in M_i are aware of sending the agent to the next stage and its not getting lost during communication. Such technique, together with a perfect determination of faults, prevents the duplication of agents in NAP. It goes without saying that an efficient determination of faults is impossible to be achieved in the Internet [1], that is why NAP cannot be always used except in controlled environments. The following approaches are not that limiting.

In Ref. [19], Rothermel and Strasser suggest an approach based on negotiations and the election of a leader, a method we have already hinted at. The agent is sent between two consecutive stages S_i and S_{i+1} using queues of negotiation messages.

More in detail, a place P_i^j puts the agent (a_{i+1}) in the queue of input messages of place P_{i+1}^k as part of a global transaction. Such global transaction matches the entire stage execution in stage S_i and includes:

- taking the agent (a_i) from the input messages queue;
- executing the agent's stage action;
- putting the resulting agent (a_{i+1}) in M_{i+1} places message queue.

All multiple executions of M_i will execute this transaction, but only the leader, selected from the election protocol, will bring them to a conclusion. All other places will abort the agent's stage action (fig. 8). Together with the use of local transactions, this approach ensures the mobile agent execution exactly-once property, but unfortunately it is vulnerable to blocking.

Such vulnerability is caused by the use of a bi-phase protocol, 2PC protocol [14]. 2PC protocol is a protocol used in distributed systems to consistently carry out computations involving several remote participants. Its goal is to indicate how to coordinate transaction participants so that it is automatically executed.

Generally speaking, 2PC takes into account the presence of a Transaction Manager (TM) that will be elected by the participant (client) that wants to carry out the transaction. Every time he wants to do a transaction, the client will have to carry out the following operations:

- communicate to TM all the participants in the transaction;
- execute all the operations forming the transaction on the involved participants;
- ask TM to carry on COMMIT operation.

  At this stage TM will have to:
- send the PREPARE command to all participants;
- wait for the participants' answers.

Two cases can occur:

- all the answers are affirmative ones (i.e. all participants can execute the operation). In which case TM sends all participants the COMMIT command;
- an answer at least is negative. In which case TM sends all participants the ABORT command, thus, provoking the transaction abortion.

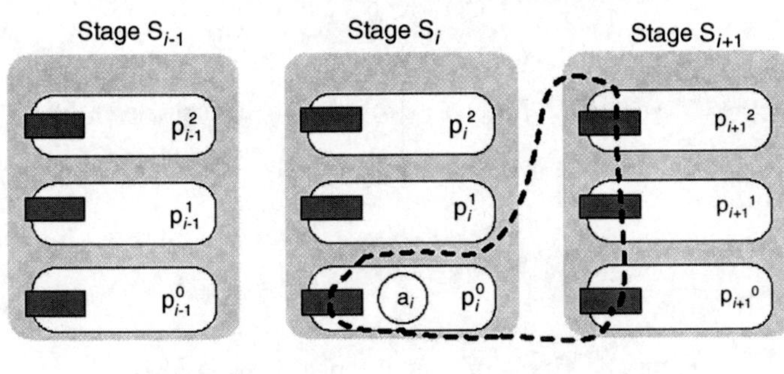

Figure 8: Stage action abortion.

On their part, participants must behave in the following way:

- receive from the client (who created the transaction) the operations and execute them;
- once received the PREPARE command from TM, establish whether the above-mentioned operations have been successfully executed (it is impossible to accept further operations from the client at this stage). On such premises TM can be answered positively or negatively;
- wait for TM's definite communication (of course if a participant answered in a negative way, it will be an ABORT).

Two cases can thus occur:

- TM communicates an ABORT. In this case operations matching this transaction will have to be "undone", bringing the state back to the previous situation;
- TM communicates a COMMIT. In this case changes are made definite and stable.

It is clear that the possibility of ABORT implies that on the participants' side there is a rollback support, which can be obtained in various ways, for example by using a log file, or making changes not on main data but on a copy, and later making them effective only after COMMIT. We can graphically represent possible participants state transitions in the following way (fig. 9):

From what we have just said we can infer that such protocol ensures exactly-once property but does not avoid the agent's execution blocking (in case of transaction *abortion*).

Assis Silva and Popescu-Zeletin [20] try to solve the problem of blocking in Ref. [19]. First a 3PC protocol is used. Since leader election protocol contributes to the problem of blocking, it is replaced by a replicated fault tolerant database, which is accessible from all places. It serves as synchronization mechanism in leader selection. This database stores, among other things, information on

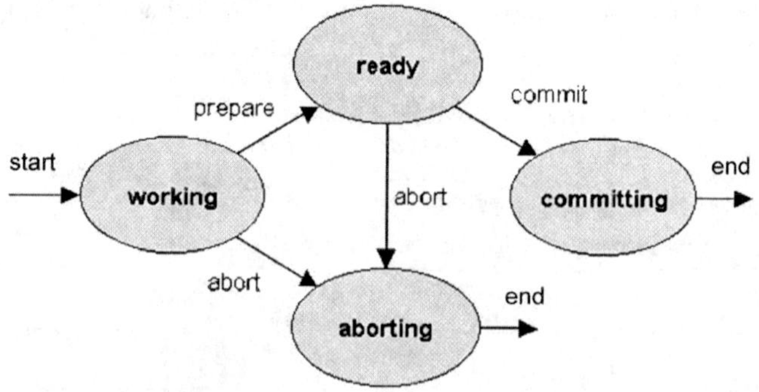

Figure 9: Graph of exactly-once states in SRB.

current leader ID. A leader can have access to this database whenever he wants and confront his ID with the current leader's one. If they are different, it means that a new leader has been selected and the old one will cancel the effect of his changes. Such database causes a considerable overhead to the agent's execution and, moreover, violates the assumption of the mobile agent's complete autonomy.

The two previous approaches are both based on a leader transaction and selection model. In Ref. [16], Pleisch and Schiper propose a simpler model, solving a problem of coordination among the places in set M_i avoiding agents' duplication. In a coordination problem (agreement problems), all participants (in our case the places in M_i) agree on a common value. In this context, places in M_i agree on:

- the place, called primary, that has been executed by the agent;
- the resulting agent (a_{i+1});
- the set of places M_{i+1} of the next stage S_{i+1}.

Primary place has concluded the stage operations of (a_i), while all other places of M_i that have only partially executed (a_i), will cancel the effect of their changes to the place state. Since all places have reached an agreement on which is the primary place, the agent's stage action will be executed just once. The entire agent's execution is, therefore, shaped as a sequence of coordination problems. Solving these agreements, at every stage of the agent's execution, prevents from agents' duplication, while place redundancy prevents from blocking. Both mobile agent's execution properties, therefore, are ensured.

In SRB approaches, the mechanisms apt to prevent from agent duplications (e.g. transactions and leader selection in Refs. [19, 20] or agreements in Ref. [16]) provoke an additional overhead to the agent's execution. In Ref. [16], this overhead depends on the agent's dimensions, while in Ref. [19], where it is based on the technique of check message exchange between the places in a stage, is independent from dimensions.

14.2 Pipeline mode

One drawback of SRB approach is the need to have a set M_i of places at every stage S_i, even when there are no faults. Such a place redundancy, thus, contributes to communication overhead between consecutive stages. The re-use of previous stage places to execute current stage improves such approaches performances and avoids high communication costs [9, 16, 21,]. Fig. 10 shows *pipeline mode* [16] with a 3rd degree replication.

At stage S_i, only place P_i is given a further destination, while P_{i-1} and P_{i-2} are used again. P_{i-2} and P_{i-1} are usually *witness* places of the execution on P_i. However, iso-places and hetero-places are also supported in this mode, though their use is limited.

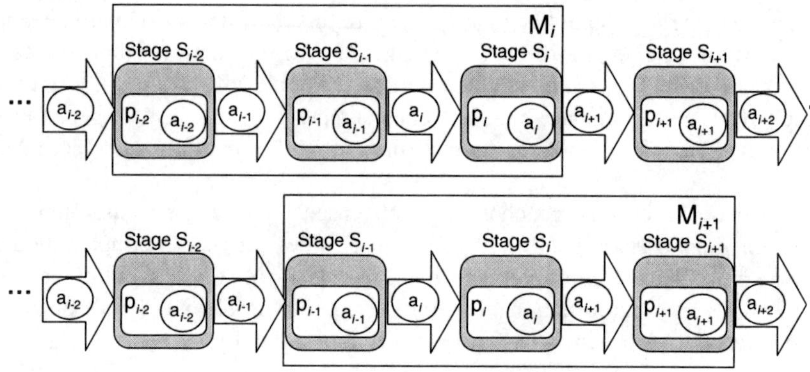

Figure 10: Pipeline mode.

15 Main differences between SRB and TRB

SRB approach uses just one kind of redundancy in stage S_i in order to prevent the agent's loss and blocking. On the other hand, TRB approach uses two different redundancy instances: (1) it stores copies of the agent (a_i) in previous places to prevent the agent loss, and (2) it executes these copies in other places P_i^I, P_i^{II},... to prevent blocking. A fault, or a wrong fault determination, in TRB approach provokes an additional sending of the agent, i.e. a fault of P_{i+1} causes the sending of (a_{i+1}) by $P_{i'}$ in another place P_{i+1}^I. This extra communication is the critical part of the agent's execution. Through other means, SRB approach does not need the additional sending of the agent. Rather the replications of the agent (a_{i+1}^j) are already available on all places of M_{i+1}. There is, therefore, an overload of communications at every stage execution, even if there are no faults.

Another important difference between TRB and SRB is the lifetime of duplicated agents. It is a crucial one, since it influences how long data items must remain *locked.* During that period, no mobile agent can have access to data items and blocks, limiting the entire system throughput.

In SRB approach, the agent's copy instances' life is limited to the stage execution. On the contrary, TRB approaches usually need to block all data items until the end of the agent's execution. When it reaches its destination, the agent's changes will be turned permanent, while all other copy agents will be determined and their actions will not be concluded (undoing). The agent's stage actions committing and undoing phases need the sending of additional messages to all places in the itinerary. That is not the case with SRB approaches.

Generally speaking, SRB approaches have high costs for stage action execution, due to overhead of communications among places of a set M_i and between two consecutive stages. The latter is reduced in the case of pipeline mode. Otherwise TRB approach must provide for the agent's stage action committing and the copy agent's one undoing. Though such overhead is not immediately visible to the agent's owner, it limits the overall system throughput.

TRB approaches, because of minor communication overhead, and unlike SRB ones, are widely used in environments where there is limited bandwidth, though with a moderate number of simultaneous agents.

16 Existing solutions

16.1 FATOMAS

FATOMAS protocol is classified among SRB approaches with agent-dependent structure. It presupposes the following hypotheses:

- asynchronous distributed network (e.g. the Internet);
- processes communicating through message passing;
- no limits to message transmission delay and to process speed;
- presence of a fault detector enabling the solution of problems concerning coordination among copies of the agent at the same stage.

Fault tolerant mobile agent execution leads to a sequence of coordination problems (agreement). Fig. 11 shows an example of mobile agent execution in four stages. It must be noticed that, on stage S_2, place P_2^0 has a fault and therefore P_2^1 must undertake the execution.

The solution of a coordination problem leads all places in M_2 to agree that P_2^1 is the place that has executed A_2.

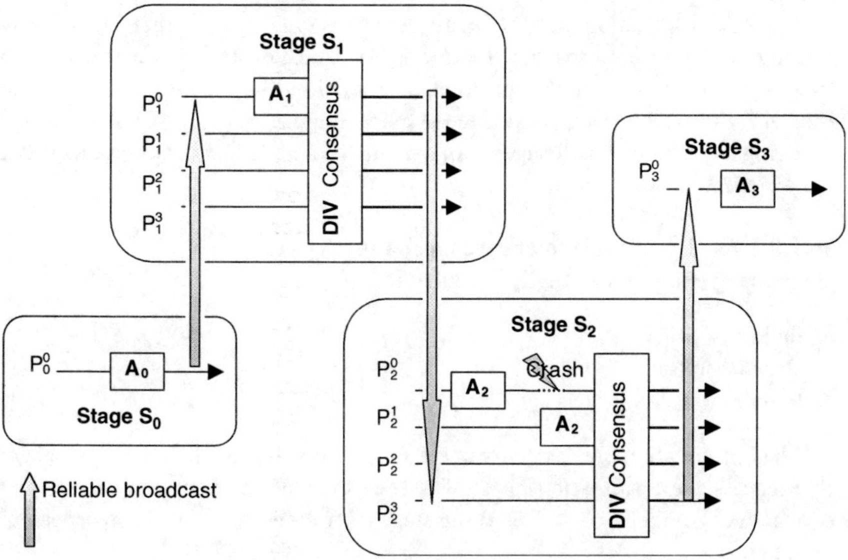

Figure 11: Execution of an agent with fault of P_2^0.

FATOMAS, being a fault tolerant mobile agent, is able to negotiate and solve two problems:

- DIV Consensus (agreement problem);
- reliable broadcast.

In every single stage S_i (1) one of the copy agents executes the stage operation, then (2) solves the coordination problem with all copy agents and, finally (3), is sent to the next stage.

Steps (1) and (2) are dealt with jointly as part of a variation of the agreement problem, called *Deferred Initial Value Consensus* (briefly DIV Consensus) [22].

In the problem of agreements (among copies of the same agents), every process needs an initial value [23] and, in the case of FATOMAS, the initial value at stage S_i for place P_i^j, is obtained from the execution of agent (a_i).

The execution of A_i on all places of stage S_i is not necessary and is too costly: in this field, DIV Consensus enables us to defer the initial value calculation of places P_i^j and make it only on demand of the DIV Consensus algorithm.

For example if P_i^0 succeeds in calculating its initial value and does not get lost (crash), no other place P_i^j, with j greater than 0, is called to do that (i.e. to calculate it).

DIV Consensus presumes that most participants do not fail. The block where DIV Consensus is implemented is fault tolerance enabler (FTE).

Step (3), instead, deals with problems concerning reliable broadcast.

A traditional reliable protocol presumes a 1–m communication scheme where a process broadcasts 1 message to m recipients. In our case, we have an r–m scheme: r senders must broadcast the same message to m receivers.

Agents' propagation occurs in an asynchronous system that, as we have already said, presupposes that there are no rigid constraints on transmission time. All that impacts on the different cases dealt with by coordination protocol (i.e. DIV Consensus) that is executed at every stage S_i.

Being asynchronous, the agent (a_i) may not be able to arrive simultaneously to the different places P_i^j of stage S_i.

16.1.1 Isolation of fault tolerance mechanisms

A mobile agent is conceptually executed in 3 phases:

a. initialization;
b. stage operation;
c. termination.

The initialization phase occurs in the source agent S_0, while the recipient agent S_n executes the termination phase. Between the source agent and the recipient one, at every stage S_i $(0 < i \leq n)$ the stage operation phase has been repeatedly executed.

Actually, fault tolerance should be independent from mobile agent and its mechanisms should be clear to the agent's owner.

Unfortunately, complete transparency is difficult to achieve and the user-defined agent, i.e. the part defining the operations peculiar to the agent, needs to interact with fault tolerance mechanisms. In the execution of an agent without copies, only the next place to reach needs to be specified; on the contrary, the execution of fault tolerant agent with copies requires (except in the particular case of pipeline mode) a set of destination places for the next stage (M_{i+1}). The agent is of course aware of the copy and complete transparency is impossible. Moreover, having to ensure fault tolerance overloads the agent's stage action with another phase: the "commit/abort" one. The presence of more copies of the agent, in fact, causes a problem of coordination that, in order to avoid overloading, must be solved through the commit/abort phase.

It has been observed that an error in detecting a fault can lead to the violation of the mobile agent execution "exactly-once" property. Such correlation problem can be solved at the source, avoiding multiple executions of the agent, by choosing a primary agent P_i^{prim}: primary carries out the operations proper to the agent, while all the other places where copy agents are being executed must execute the "abort/undo" operation, i.e. must finish and/or cancel all operations executed till then. "Commit/abort" phase semantics stems from the nature of the agent's operations: database transactions need to be "committed" or "aborted" depending on their being carried out by the agent P_i^{prim} or by one of its copies, while idem-power operations (whose repetition does not causes unwelcome effects) usually do not need further measures.

The architecture incorporating fault tolerance mechanisms in a single component called "fault tolerance enabler" (FTE) is hereafter reproduced.

Fig. 12 shows the interaction flow between FTE and the user-defined agent: this interaction occurs during the phases of stage operation "commit/abort".

FTE groups together the mechanisms managing fault tolerance, while the "user-defined" agent contains the application-specific part.

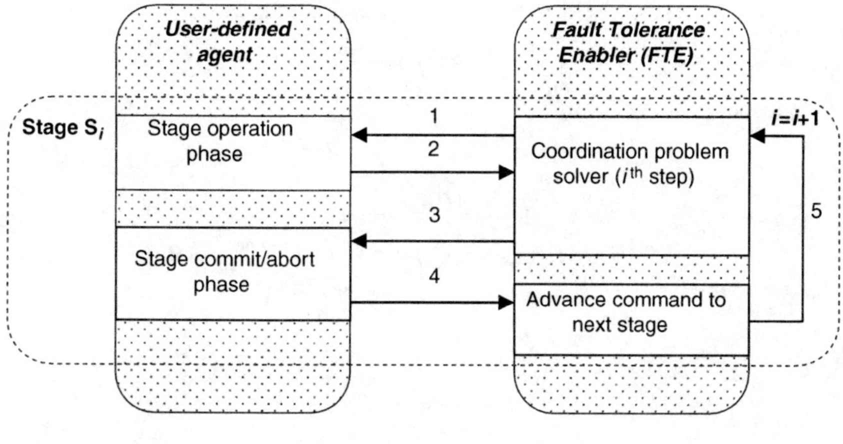

Figure 12: Phase of a mobile agent execution and of interaction with FTE.

At every stage S_i ($0 < i < n$), FTE solves the problem of coordination among stages: depending on coordination result, the operations achieved during stage operation phase (fig. 12, arrows 1 and 2) are "committed" or "aborted" (arrows 3 and 4).

Finally, FTE shifts the agent to the group of places in M_{i+1} (arrow 5) that are calculated by the user-defined agent and returned as the result of stage operation phase.

As to the place where the FTE block is located, it is possible to identify two different approaches:

a. *agent-dependent* approach: FTE travels with the agent;
b. *place-dependent* approach: FTE is located in places.

16.1.2 Agent-dependent approach

In the agent-dependent approach, used by FATOMAS, FTE block is integrated into the agent and travels with it. Every agent has just one instance. FTE is initialized by the user-defined agent in the *agent source* and causes the user-defined agent termination phase in the *agent destination.* Interaction between a user-defined agent and FTE generates a fault tolerant mobile agent. Replica mechanisms are, therefore, completely transparent in places: the agent appears in the place as a normal agent.

Consequently, existing mobile agent platforms do not need to be changed: instead of programming the agent according to the API platform that owns it, the agent makes use of FTE–API's functionalities (fig. 13). FTE then manages problems such as those concerning fault tolerance and mobility: for example in mobile agent platform ObjectSpace's Voyager [24], an agent moves after a call to "move" method. Together with destination, the "move" method also needs the name of the method that must be called to reach destination place.

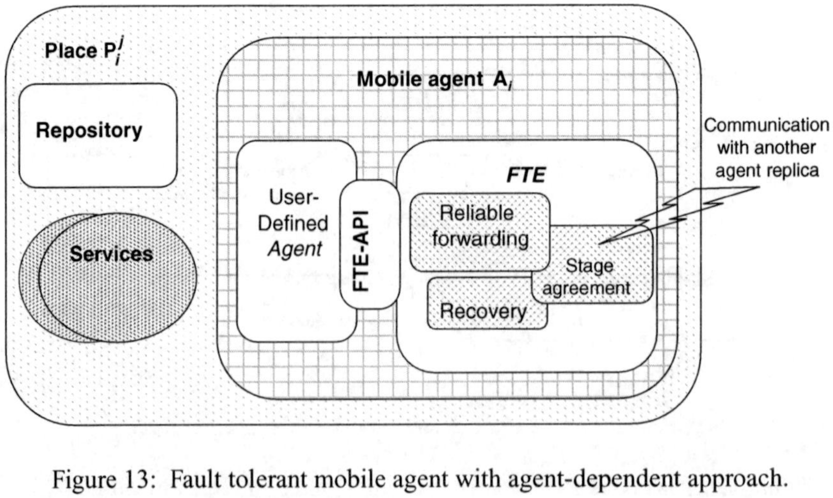

Figure 13: Fault tolerant mobile agent with agent-dependent approach.

In FATOMAS, the method called *doStageOperation,* is invoked by FTE−API every time the agent reaches a new place. The results of such method are the addresses of the next destination places.

Fig. 13 shows the architecture of agent-dependent approach: FTE is composed of a block called *stage agreement* (dealing with agreements among copies and implementing DIV Consensus algorithm), a *reliable forwarding* block (responsible for the agent's moving to next stage), and a block called *recovery* dealing with the agent recover in case of fault or in case it arrives late to a place.

Last, there is a location called *repository*, where information concerning fault tolerance can be temporarily stored. This location depends from the agent's platform, but is like a local storeroom, such as the Voyager directory [24]. *Repository* is also used to receive in place P_i other agents located in other places P_k (with k different from i) that must have remote access to such information.

16.1.3 Place-dependent approach

Here, FTE is at the mobile agent's disposal [17, 18]; thus fault tolerance is included in places, and FTE instances are created and executed at every stage of the mobile agent's execution.

A disadvantage of place-dependent approach is linked to the fact that it is necessary to modify the mobile agent's owner platforms; in particular, the existing basic platforms must be replaced by platforms including all fault tolerance mechanisms, and that is problematic. Moreover, providing locally to fault tolerance mechanisms can lead to homogeneity problems.

An advantage of place-dependent approach is that it assigns to the agent A_i^j's copies, if necessary, the places where they will operate. Actually, only the copies of the agent whose stage operation phase has been really executed are instanced. In every place it is executed one instance of FTE (a different one for each copy A_i^j of the agent).

Furthermore, the fact that FTE is situated in places implies that it must not be transported with the agents and, therefore, their dimension is limited, to the benefit of a better performance in transmissions.

In favour of agent-dependent approach, on the contrary, there is the fact that if two agents A and B must execute stage S_i in the set of places M_i, FTE can be re-used without any change. However, the gain in terms of performance is not significant.

16.2 James

In contrast to FATOMAS, we are now going to describe how the James system, classifiable as TRB place-dependent, manages fault tolerance.

In James project [25], a mobile agent infrastructure based on Java has been developed with an advanced support for network managing.

The environment for mobile agent execution provided by James platform is thus structured:

a. there is a platform *agency* in every host of the network. They provide the necessary mechanisms for agent migration;
b. there are one or more *agent managers*, *entry point* nodes of the system, enabling the launch in the network of applications based on mobile agents and provide for the management and monitoring of agents execution.

User applications, in James platform, are written in Java and make use of James API [26] for mobility control. After having written an application, the programmer must create a *jar* file that must be put in the Code Server. Host machines executing James manager provide a graphical user interface (GUI) for the remote control and monitoring of agents, agencies and applications. With this interface the user can manage all system's agents and agencies, observing their execution state, starting them, and suspending or installing new agents and agencies.

James platform has been created with a particular concern for fault tolerance and system's strength problems.

16.3 MESSENGERS

MESSENGERS is a distributed system based on autonomous object called precisely "Messengers". It partitions the network in three levels (fig. 14):

- *Physical network* is the basic resource for computations.
- *Daemon network* is an ensemble of server processes whose task is to interpret messengers functions and system commands (an example of system command is

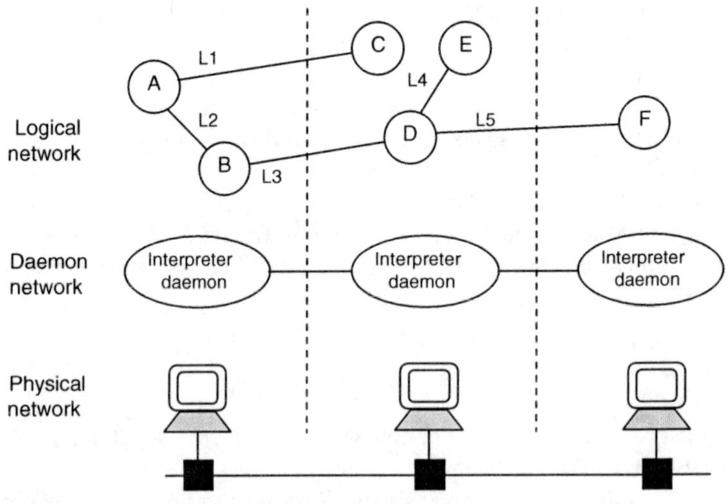

Figure 14: The MESSENGERS system.

the one concerning Checkpointing beginning). Each physical node corresponds a daemon. In publications *daemon* is intended as a program working on the background and achieving services in a repetitive way, as in a vicious circle. This term is typical of Unix environments, while other definitions are used with other operating systems, for example the server one. Traditionally, most daemon programs have a name ending by the letter «d».

- *Logical network* is run-time created over the daemon network. The logical network great number of nodes can be created on the same nodes of the daemon network, which runs on physical network's nodes. Logical nodes can be interconnected by arbitrary typology logical links. Every logical network connection has a name and a different weight that messengers uses to determine which one to use. Every logical node has a name and provides a memory space accessible to all messengers gathering there. This memory space, called *node variable area*, function both as a database and as a communication channel among messengers.

Every messenger can have access to three kinds of variables: messenger, node and environment variables. Messenger variables are local ones and are carried by the messenger that spreads them throughout logical network. Node variables, residing in the node, are mapped in the node variable area where the messenger currently in execution and are shared by messengers running in the same node.

Environment variables give information such as the current node's name, or the last link's name and weight.

Messengers navigates in logical network through explicit navigation statements, enabling also the creation or destruction of logical links and/or nodes. Messengers can also carry out computations of various kinds in the nodes they visit: they can contain computational statements, or can invoke ordinary functions in C. Other information on MESSENGERS can be found in Ref. [27].

16.3.1 Daemon's state capture

When no messenger is in execution, the daemon state is easily detectable: it consists in a data space shared by messengers (the logical network) and by the ensemble of messengers in execution queue or waiting for the input/output phase. Every messenger is represented by a simple data structure, called *messenger control block*. Moreover, when the agent reaches the daemon or prepares to migrate, its state is captured in such data structure.

Techniques for the agent state capture have been implemented in Odyssey [28], IBM Aglets [29], Ara [30], Agent Tcl [31], TACOMA [7] and other mobile agent systems.

16.4 Configurable mobile agent

This solution [32] considers a mobile agent migration algorithm and a fault tolerance management one. Actually, it manages fault tolerance cases in which the agent cannot migrate to a server or cannot "work" in a server. The complete fall of a server in which a mobile agent operates is not managed.

The algorithm in question stems from the remark that it is often the entire code of the mobile agent to be transferred from a server to another, whereas actually only a part of it is needed in a particular host. It has been thought, therefore, of "decomposing" the agent into various parts and sending every host where the agent must migrate only the part of code actually useable, with a consequent decrease of load for the network. In a sense, the agent will be able to operate in an almost completely parallel way.

16.4.1 Structure of the agent

According to this model, the mobile agent is composed of the following elements (fig. 15):

a. *Kernel*: it performs the scheduling of function modules on the basis of FSLM (formal schedule list of modules) description;
b. FSLM (fig. 16);
c. *Message and data buffer*: buffer destined to the exchange of messages and data;
d. *Function modules*: modules each of which represents a typical agent's functionality.

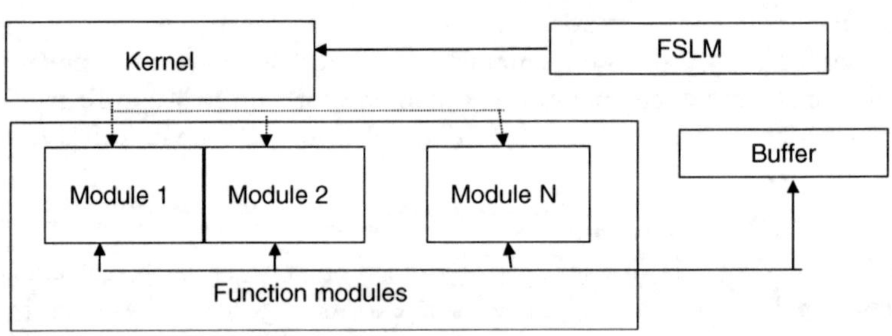

Figure 15: Agent structure in MESSENGERS.

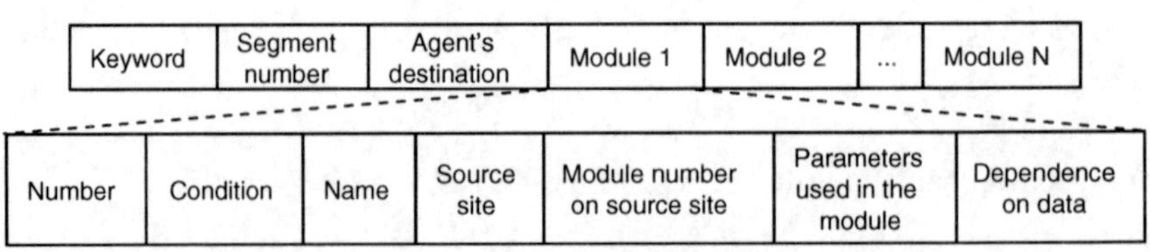

Figure 16: Segments structure of FSLM.

16.5 ACS

This method takes into consideration the presence of a server managing mobile agent creation in order to relieve the user of such burden. This server is called ACS (Agent-Creating Server). It includes a library of function modules (ML) that is a collection of all mobile agents function modules; a mobile agent service module (MASM) that receives the users' requests, interprets them and, if there are no grammar errors, creates the agent from library ML; a record list of all FSLM (RLFSLM), an agent record list (ARL), an agent launcher (AL), and, finally, a GUI.

16.5.1 Migration of an agent

The method proposes an alternative process of agent migration in order to speed and do not overload the network (fig. 17). The starting assumption is that all agent

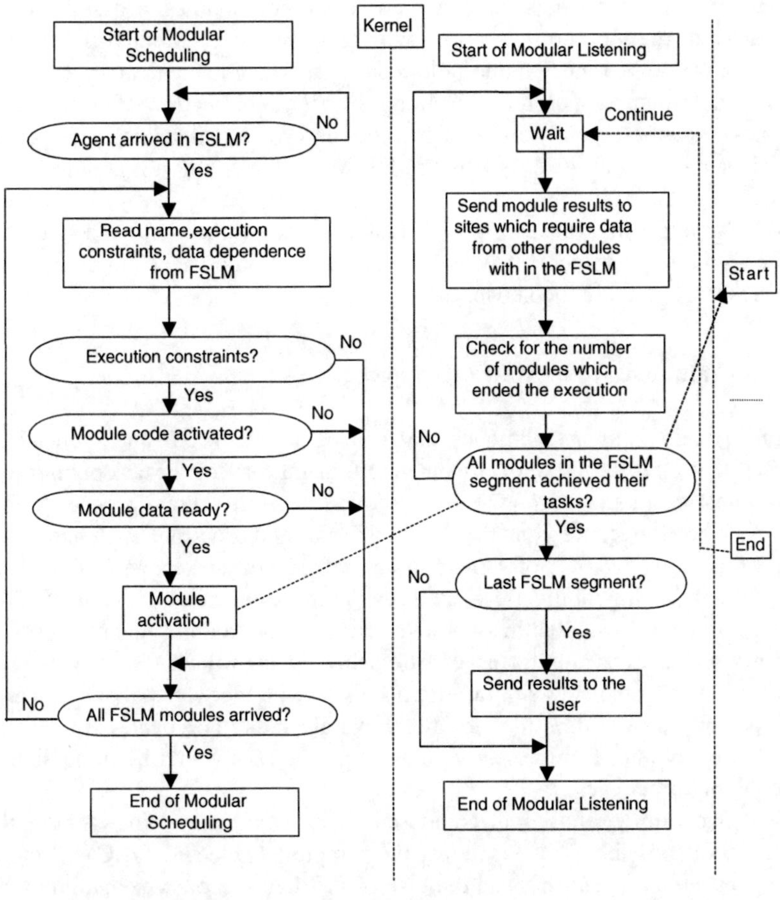

Figure 17: Migration algorithm flowchart.

destinations are *a priori* known; the launcher, in fact, sends the agent's kernel and FSLM to all destination servers, after having taken the agent number from ARL and its FSLM from RLFSLM. Then the launcher analyses FSLM and sends the necessary modules to each destination. If modules are already present on the local ACS, the launcher sends them directly, otherwise it asks for them to the site where they are. In any case, after having sent such modules, the launcher waits for a system confirmation message. Only after having received all confirmations it will delete the agent from ARL. During transmission a certain security is assured by the use of Concordia [33] protection procedures; the agent's codes and data are in fact encrypted. For further reliability, moreover, ARL and ELFSLM data are stored in non-volatile memories. The kernel contains moreover two threads called ModularScheduling and ModularListening.

ModularScheduling waits for the agent's FSLM. When it arrives, it reads from FSLM the module's names, their execution conditions and the list of data dependences. Only when execution conditions have been verified, and contemporarily the modules' codes and possible other modules' necessary results have arrived, the module in question will be activated by ModularScheduling. The process will be repeated until all modules in FSLM are activated.

The main function of ModularListening is, on the contrary, that of inspecting all modules' execution processes and, when it finds a module that has accomplished its task, sends data to the sites listed in FSLM. At the end of all modules it will show all results.

ICMP monitoring package detects faults in communications and is strictly linked to TCP/IP protocol. It is used to signal error messages but also to find specific information in the network.

16.6 Technique based on ICMP packages

Among widely known techniques that can be found in current publications, another solution is the use of intelligent mobile agents having the task of migrating from a host to another throughout the network ready to signal a possible fault [34]. An algorithm concerning this technique has been proposed by Puhan Zhang and Yufang Sun from the Software Institute of the Chinese Academy of Sciences. It exploits an intelligent mobile agent through the monitoring technique of ICMP package, implementing that in Java on the IBM's Aglet mobile agent system.

While exhaustive monitoring can be achieved in little LAN network, it is unthinkable to implement similar algorithms in wide networks. The number of packages in fact would be too elevated. Several faults, moreover, could coexist simultaneously and, from excessive flow of packages it would be impossible to trace their number back.

The algorithm proposed is composed of a fixed monitoring system that locally stores possible faults, isolating the symptoms detected by ICMP, and an intelligent mobile agent that has the task of visiting all nodes exchanging with them information concerning faults.

The entire architecture presented in this project takes into account the distinction between a server Manager and the ordinary ones called Network Component (NC). While in Manager a static agent called Ad and a dynamic one called Max are both present, in every NC there is an intelligent mobile agent indicated as IMA.

IMAs monitor ICMP packets and discover faults locally. Instead of establishing client−server connections in order to gather the necessary information to execute fault detection, they process fault detection locally, detecting a simple fault and isolating the single symptom from the ICMP package observed. Max's goal is to visit all nodes containing information about any fault.

Fault detection by an IMA occurs according to the following steps:

- the manager provides every IMA with an ICMP package and sends it to all present NCs;
- IMA isolates in every NC the symptoms coming from the observation of packages with not reached destination;
- IMA traces a simple fault on the basis of these symptoms diagnosis;
- in order to solve fault problems, IMA must know all detected faults. Besides examining locally detected symptoms, it receives therefore from the manager database all information concerning other nodes' faults;
- IMA sends the isolated symptoms to the manager that will use them for fault diagnostic and detection;
- the manager diagnoses all discovered symptoms and faults and sends them to IMA;
- the manager creates a mobile agent Max that migrates to all nodes needing gathered information.

Intelligent agent is made up of two parts: the *Packet Monitor* and the *Symptom Isolation*. The *Packet Monitor* does nothing but distinguish and catalogue the encountered kinds of fault in three common typologies: Echo Request, Echo Reply and Destination Unreachable. It also separates and arranges unreachable destination addresses. Each unreachable destination is a symptom. Symptoms are arranged in a *symptom table* according to their number of occurrences. *Symptom Isolation* uses such table to isolate symptoms and establish if it is possible to go back to a fault. If that is possible, symptom isolation sends local results to fault database and to the manager Ad. Finally, the manager will reconstruct the complete fault on the basis of received data, and it will solve the problem if it is able to.

16.7 MATS

MATS [35] is a system that extends the concept of parallel virtual machine (PVM) architecture, using a combination of collaborating mobile agents to obtain automatic and dynamic configuration of distributed processes. MATS is implemented in Java, using the ObjectSpace Inc.'s Voyager [24] mobile agent platform that makes use of Java object serialization. PVM and other similar systems try to use the potential offered by modern workstations connected by fast networks, combining distributed calculation resources in a single virtual computer, usually indicated as distributed virtual machine (DVM). Ideally, the complexity of how, where and why processes are executed should be hidden by high-level interface.

A possible way of using mobile agents for DVM systems project is to create a generic agent, residing in each machine and checking local resources, equivalent to PVM daemon process. This is a very common approach, although it has the disadvantage of creating a relatively big and complex agent not always able to manage a great number of functions, such as

- *navigating*: maintain an itinerary and the knowledge of target sites;
- *negotiation*: ability to negotiate services and payments with host servers;
- *security*: protection against malicious attacks from server hosts or agents;
- *task management*: ability to achieve the assigned task;
- *communication*: ability to establish communication with other agents;
- *user interface*: interact with the user through graphic interface.

All these characteristics, though advisable, contained in a single agent, contribute to widen its dimension.

MATS, instead of integrating in a single agent all necessary functions, implements various kinds of agents having different functions and abilities. Since these agents must cooperate as a team, the system is called MATS (mobile agent team system). One of MATS's projected goals was the minimization of the code quantity that needs to be moved through the network. That has been done localizing the most complex functionalities in a non-mobile agent that can communicate with simple specialized mobile agents through a mechanism of message passing.

The ensuing mobile agents are lightweight, i.e. they require less processor time to be serialized, and can be transmitted more rapidly. That finds an important implication in parallel application efficiency. If the agents were too big and required many machine resources to be transmitted, the benefit of solving the problem in parallel would be lost.

16.8 A³

A³ is a framework supporting mobile agent distributed calculation. A³ indicates three different fault tolerance goals:

Access security: the idea is that if an agent does not come from the same community, which can be either a village or a public Intranet (PI), it is denied access to common resources;

Agent persistence: the aim is to enable a mobile agent to go away from a dangerous area subject to possible failures. Possible malfunction previsions can be done by calculating some reliability indexes based on a set of parameters of the execution system;

Avoidance of long queuing time for service: an agent should move from a site requiring a long queuing time for service. Since that kind of migration (from an overloaded site to an idle one) produces load redistribution among sites, it is a feature introducing load dynamic balancing ability in framework A^3.

The main distributed programming environments (DPE), toward which A^3 is directed, are two:

- *function-oriented DPE*: in this paradigm the components forming the distributed program are objects or agents that can be separated and distributed as entities encapsulated to different nodes for parallel execution. The advantage is to obtain distributed parallelism speed and to improve reliability thanks to components' redundancy. Such reliability is necessary when it comes to distributed applications with critical timing, which should not be interrupted once begun;
- *data-oriented DPE*: this kind of environment's objects or agents are sent towards places where great amount of data are. The agents bring back their computation results only after having processed data. This approach reduces network traffic and is, therefore, cheaper than RPC traditional approach.

16.8.1 Definition of a community

The smallest community supported by A^3 is a village that can be connected with other villages to form a wider community called PI. An Internet node, however distant, can voluntarily enter and be part of a village or leave it, through a membership registration or deregistration system. Only the members of a community can share common resources that are on the contrary denied to non-members.

16.8.2 Migration indexes

A^3 agent migration has two goals: the former is to obtain *agent's persistency* on the basis of reliability indexes, the latter is to obtain *dynamic load balancing* so as to avoid long queuing time for every remote server service.

Agent's persistency: Migration depends on dynamically calculated values of the following reliability indexes:

- *Load reliability index* (*LRI*): a host calculates periodically its own reliability index (LRI) and records it in local NCT (network configuration table). If LRI's current value is lower than a threshold fixed in advance, the host will try to make all local mobile agents migrate to another safer site.
- *Internal reliability index* (*IRI*): a mobile agent calculates current host's IRI value, to decide if it is reliable or not. If such value is lower than a threshold fixed in advance, the agent asks the host to start its migration to a safer site.
- *External reliability index* (*ERI*): it is an index calculated by the agent that has created another mobile agent, on the basis of some parameters of the latter's

current host. If ERI's value is lower than a threshold fixed in advance, the creator agent can ask the "child" agent to move in the present host. However, final decision is up to the child agent that compares current values of LRI, IRI and ERI.

Dynamic load balancing: In order to enable agents' autonomous migration from overloaded nodes to idle ones, the following indexes are introduced:

- *Local migration index* (*LMI*) – it is calculated by the host on the basis of the following execution system parameters: (i) mean number of queuing service demands, (ii) use of main memory, (iii) speed of new queuing demands, (iv) speed of memory decreasing availability, (v) frequency of I/O accesses to peripheral devices and (vi) threshold values stemmed from past operations.
- *Internal migration index* (*IMI*) – a mobile agent evaluates the answer time needed for a service according to (i) average queuing waiting time in local system, (ii) assessment of time required by the service, and (iii) amount of required main memory.

The main goal of A^3 is to give distributed computations access security, as well as agents' persistency and to avoid long queuing times to obtain a service through load dynamic balancing. At prototypal level, it has been carried out using Aglet mobile agent platform. It has been tested on the Internet, with a selection of sites that includes Hong Kong University Intranet and some other local Internet sites. The next step is the prototype test in more dynamic Internet environments that include different remote and proxy Internet sites.

16.9 FLASH

FLASH (FLexible Agent System for Heterogeneous Cluster) [36] system can be used to create complex parallel applications without any detailed knowledge of underlying cluster. Mobile agents deal with the execution of these applications so as to get the cluster's nodes' load balancing and fault management in a transparent way to the user. Fault tolerance mechanisms implemented by FLASH are based on those adopted by FANTOMAS (FAult toleraNT apprOach for MObile AgentS).

16.9.1 The FANTOMAS model

This model [37] is based on a checkpoint technique for fault tolerance management. In order to carry out such model, a modular structure of the agents is implied as well as the users' possibility of enabling or not fault tolerance. For each mobile agent called *user agent* (UA), for which fault tolerance activation is required, a *logger agent* is created, thus forming a couple of agents.

The logger does not take active part in applications execution and, therefore, needs only a slight use of CPU. It follows the UA but never resides in the same node, so that they cannot fall down in case of a fault in the same node. The two agents keep constantly in touch through messages so that, if a fault is detected by one of them, the latter can rebuild the former by using local information. The creation of a couple of agents stems directly from migration. In order to create a logger agent, the UA serializes its own state, as it happens in migration, and sends it to a proper place. Here, de-serializing the received data creates the new agent. The difference between a standard migration and this one is that the new agent does not start with the execution of applications modules, leaving the UA free to achieve normally its own tasks. Checkpoints strategy in fault tolerance unit consists in the choice of the right moment to carry out checkpoint saving. Moreover, making this process a periodical one, migration is facilitated, because it involves the agent's serialization state.

This approach supports fault tolerance against nodes and agents' fall, as well as for communication and distributed applications. It does not need the agent's prearranged itinerary.

Following the concept of agent, diagnosis should be made autonomously and cooperatively. Transparency to the user, usually applicable, implies that the agent's fault tolerance unit does not have any knowledge of the application, especially about the meaning of results, and, therefore, cannot carry out any corrections. In order to exploit the knowledge of applications and enable the search for wrong computation results, modularity enables the additional option of application test acceptance, if the programmer has calculated that in advance. All that can be done through an interface between the agent's kernel and in turn the fault tolerance unit, the application modules, the tests to be made and the results provided by fault tolerance unit. It is possible that the standby logger is modified by a difference in the obtained results, or in confronting a message. Supposing that occurs in a transparent way to the user, non-dependence of information is required as far as their location is concerned, since it must be contained in the final results.

16.9.2 Types of agents

The FLASH system is composed of four basic components. The system agents (SAs) are three, node and user are organised hierarchically. The fourth basic component is an interface to the different services of the system (fig. 18).

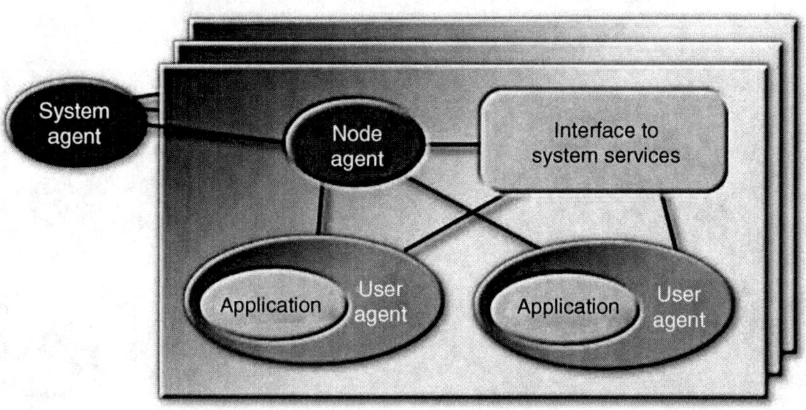

Figure 18: Components of the FLASH system.

At the top, the SA watches the cluster and keeps updated information about its nodes. If a relevant change occurs in the system, the relating (SAs) will distribute that piece of information to the dedicated node agents (NA). NA are permanently located on each cluster system node. They gather and keep information on the UAs locally residing. In addition, they provide connections to services system and can support decisions on migration. Since migrating UAs need a well-known communication partner in order to obtain local information after migration, node's agents are permanent. For dedicated tasks needing mobility, nodes agents can create messenger agents. They achieve a given task on behalf of parent's NA. UA are responsible for parallel application load balancing. Therefore, they migrate through the cluster system in search of free resources. Their decisions concerning migration are based both on inner states and on outer ones, for example on interactions with other agents, with the services system, or with local NA. There are two possible ways to include the user's application in FLASH system.

In the first one, the UA itself includes a part of the parallel application and executes it inside a thread. In the second one, the user can create a native services system offering a specific functionality for the task that must be achieved. In both cases, the only thing the programmer must implement is the functionality of the user program (or of the service system). All other system components execute their operations transparently to the user and are application-independent.

Fig. 19 shows the UA's inner structure. Besides the application thread, there exist further modules responsible for migration, communication and load management. In addition, the UA contains a data space shared among the threads. The empty module shown in fig. 19 means that other services can be added in the future, such as a module for fault tolerance. The thread to manage threads is responsible for load balancing. The module makes use of two possible sources of information. It can gather local or global information on the system load by calling a special system service. Alternatively, the module has an interface with the application thread made through a special method. This method can – but not necessarily

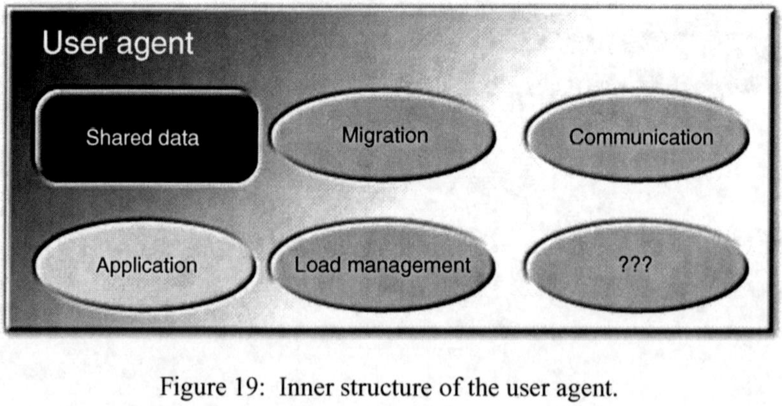

Figure 19: Inner structure of the user agent.

must – be implemented by the user and gives back an estimate of the application remaining resources, e.g. the CPU using time. If estimate is not possible or is not desired by the user, load balancing is carried out exclusively on the basis of the system already known information. Otherwise, the above-mentioned interface can be used by the application thread in order to obtain and apply the system load information for an application-integrated load balancing. Consequently, FLASH combines in a single environment load balancing implemented on the system with the one integrated in the application. The combination shows a great number of benefits. For example FLASH is able to efficiently react to load variations.

16.9.3 Service manager

Service manager keeps a system of active services and supports agents for their use. FLASH enables the programmer to extend the system with other services.

16.9.4 Fault tolerance

FLASH imports FANTOMAS fault tolerance mechanisms. It is, therefore, able to overcome single agents or nodes malfunctioning. It assumes that the agents never send wrong messages. A single agent can have a crash because of a hardware or software fault. Nodes can be momentarily or permanently out of order, also due to a hardware or software fault. When a node is malfunctioning, all agents in execution in that node are involved.

All that leads to the consideration that the diagnosis of malfunctioning should be made autonomously. The application programmer can provide some acceptance tests to ensure the UA's correct operation. The NA can detect the local UAs crash and monitor remote nodes with alive and timeout messages.

FLASH can be used to create complex parallel applications, fault tolerance being considered accessory. If required by the application, it can be introduced transparently to the user. The approach followed is similar to the primary/backup model, with the difference that the backup agent does not replace the primary one in case of malfunctioning, rather it recreates it. Such functions integration in mobility primitives facilitates the programmer's task.

References

[1] Fischer, M., Lynch, N. & Paterson, M., Impossibility of distributed consensus with one faulty process. *Proc. of the Second ACM SIGACT-SIGMOD Symposium on Principles of Database Systems*, Atlanta, Georgia, pp. 1–7, 1983.

[2] Kliger, S., Yemeni, S., Yemini, Y., Ohsie, D. & Stolfo, S., A coding approach to Event Correlation. *Fourth International Symposium on Integrated Network Management*, 1995.

[3] Mansfield, G., Ouchi, M., Jayanthi, K., Kimura, Y., Ohata, K. & Nemoto, Y., Techniques For Automated Network Map Generation using SNMP. INFOCOM '96, March 1996.

[4] Stallings, W., *SNMP, SNMPv2, SNMPv3, and RMOM 1 and 2*, Addison-Wesley: Reading, 1999.

[5] Perkins, D.T., *RMON Remote Monitoring of SNMP-Managed LANs*, Prentice Hall: Englewood Cliffs, NJ, 1999.

[6] Johansen, D., van Renesse, R. & Schneider, F.B., Operating system support for mobile agents. *Proc. of the 5th IEEE Workshop on Hot Topics in Operation Systems*, Orcas Island, USA, 1994.

[7] Johansen, D., van Renesse, R. & Schneider, F.B., An Introduction to the TACOMA Distributed System, Version 1.0. Report. Institute of Mathematical and Physical Science, Department of Computer Science, University of Tromsø, Norway, 1995.

[8] Minsky, Y., van Renesse, R., Schneider, F.B. & Stoller, S.D., Cryptography support for fault tolerant distributed computing. *Proc. of the 7th ACM SICOPS European Workshop*, ACM Press: Connemara, Ireland, pp. 109–114, September 1996.

[9] Strasser, M. & Rothermel, K., Reliability concepts for mobile agents. *International Journal of Cooperative Information System* (*IJCIS*), **7(4)**, pp. 355−382, 1998.

[10] Baumann, J., Hohl, F. & Rothermel, H., Mole − Concept of a Mobile Agent System, Technical Report TR-1997-15, Universität Stuttgart Fakultät Informatik, Germany 1997.

[11] Vogler, H., Kunkelmann, T. & Moschgath, M.L., Distributed transaction processing as a reliability concept for mobile agents. *Proc. 6th IEEE Workshop on Future Trends of Distributed Computing Systems* (*FTDCS'97*), IEEE, 1997.

[12] Murphy, A.L. & Picco, G.P., Reliable Communication for Highly Mobile Agents, Report (WUCS-99-15), Washington University, St. Louis, 1999.

[13] Milojicic, D.S., Guday, S. & Wheller, R., Old wine in new bottles, applying OS process migration to mobile agents. *Proc. of the 3rd Workshop on Mobile Object System, 11th European Conference on OOP*, June 1997.

[14] Bernstein, P., Hadzilacos, V. & Goodman, N., *Concurrency Control and Recovery in Database Systems*, Addison-Wesley: Reading, MA, 1987.

[15] Schneider, F., Towards fault-tolerant and secure agentry. *Proc. of the 11th International Workshop on Distributed Algorithms*, Saarbrücken, Germany, 1997.

[16] Pleisch, S. & Schiper, A., Modeling fault-tolerant mobile agent execution as a sequence of agreement problems. *Proc. of the 19th IEEE Symposium on Reliable Distributed Systems* (*SRDS*), Nuremberg, Germany, pp. 11−20, October 2000.

[17] Mohindra, A., Purakayastha, A. & Thati, P., Exploiting non-determinism for reliability of mobile agent systems. *Proc. of the International Conference on Dependable Systems and Networks*, New York, pp. 144−153, June 2000.

[18] Silva, L., Batista, V. & Silva, J., Fault-tolerant execution of mobile agents. *Proc. of the International Conference on Dependable Systems and Networks*, New York, pp. 135−143, June 2000.

[19] Rothermel, K. & Strasser, M., A fault-tolerant protocol for providing the exactly-once property of mobile agents *Proc. of the 17th IEEE Symposium on Reliable Distributed Systems* (*SRDS*), West Lafayette, Indiana, pp. 100−108, October 1998.

[20] Assis Silva, F.M. & Popescu-Zeletin, R., An approach for providing mobile agent fault tolerance. *Proc. of the Second International Workshop on Mobile Agents* (*MA*), *LNCS 1477*, eds. K. Rothermel & F. Hohl, Springer-Verlag: London, UK, pp. 14−25, 1998.

[21] Johansen, D., Marzullo, K., Schneider, F.B., Jacobsen, K. & Zagorodnov, D., NAP: Practical fault-tolerance for itinerant computations. *Proc. of the 19th IEEE International Conference on Distributed Computing Systems* (*ICDCS*), Austin, Texas, June 1999.

[22] Défago, X., Schiper, A. & Sergent, N., Semi-passive replication. *Proc. of the 17th IEEE Symposium on Reliable Distributed Systems* (*SRDS*), West Lafayette, pp. 43−50, 1998.

[23] Chandra, T. & Toueg, S., Unreliable failure detectors for reliable distributed systems. *Journal of the ACM*, **43(2)**, pp. 225−267, 1996.

[24] ObjectSpace. *Voyager: ORB 3.1 Developer Guide*, 1999, available at http://www.objectspace.com/products

[25] Silva, L.M., Simões, P., Soares, G., Martins, P., Batista, V., Renato, C., Almeida, L. & Stohr, N., JAMES: A platform of mobile agents for the management of telecommunication networks. *Proc. of IATA'99*, Stockholm, Sweden, August 1999.

[26] http://james.dei.uc.pt

[27] http://www.ics.uci.edu/~bic/messengers/messengers.html

[28] Odyssey API, available at http://www.genmagic.com/technology/odyssey.html

[29] Venners, B., Under the hood: The architecture of aglets, JavaWorld, April 1997, available at http://www.javaworld.com/javaworld/jw-04-1997/jw-04-hood.html

[30] Peine, H. & Stolpmann, T., The architecture of the Ara platform for mobile agents. *Proc. of the First International Workshop on Mobile Agents MA'97,* Berlin, Germany, April 1997, available at http://www.uni-kl.de/AGNehmer/Projekte/Ara/index_e.html

[31] Gray, R., Kotz, D., Cybenko, G. & Rus, D., Agent Tcl. *Itinerant Agents: Explanations and Examples*, eds. W. Cockayne & M. Zyda, Manning Publishing, 1997.

[32] Wu, C., Liu, S., Wang, B., Shi, Z. & Gu, H., Configurable mobile agent and its fault tolerance mechanism, *Proc. of the 2001 International Conference on Computer Networks and Mobile Computing*, p. 380, 2001.

[33] Walsh, T., Paciorek, N. & Wong, D., Security and reliability in Concordia. *31st Annual Hawaii Int. Conference on System Sciences*, HCISS31, January 1998.

[34] Zhang, P. & Sun, Y., A new approach based on mobile agents to network fault tolerance, *Proc. of 2001 International Conference on Computer Networks and Mobile Computing*, Beijing, China, pp. 229–234, 2001.

[35] Ghanea-Hercock, R., Collis, J.C. & Ndumu, D.T., Co-operating mobile agents for distributed parallel processing, ACM − Autonomous Agents'99, Seattle, 1999.

[36] Gonne, M., Grewe, C. & Pals, H., Monitoring of mobile agents in large cluster systems, IEEE, 2001.

[37] Pals, H., Petri, S. & Grewe, C., FANTOMAS: Fault tolerance for mobile agents in clusters, *Proc. of 15 IPDPS 2000 Workshops*, Cancun, Mexico, eds. Rolim, J. et al.: Parallel and Distributed Processing, number 1800, in LNCS, Springer-Verlag: Berlin, pp. 1236–1247, 2000.

Security in mobile agent systems

Pietro Alessi, Agostino Bellavia, Marco Biffarella, Giorgio Borelli, Giuseppe Bosco, Agostino Buono, Roberto Caico, Lorenzo Cassar, Giuseppe Ciulla, Marco Conti, Calogero De Gregorio, Demis Di Rosa, Giuseppe Ferrigno, Giuseppe Francaviglia, Mario Giurlanda, Giuseppina La Fiura, Francesco Lo Piccolo, Martino Maggio, Giuseppe Davide Mannone, Fabio Pisciotta, Paolo Ponente, Giovanni Reina, Salvatore Sorce, Antonino Tamburello and Alessandro Genco
DINFO – Dipartimento di Ingegneria Informatica
Università degli Studi di Palermo

1 Security in the network

In this section, we will analyse some security aspects that must be ensured in an Internet/Intranet environment and, generally speaking, in any network communication environment. Significant aspects and problems for our purposes can be briefly summed up as follows:

- *authentication*: the first important problem having to cope with in terms of security is to verify that the one that has access to certain network resources is actually the one he claims to be.
- *access control*: another aspect strictly linked to correct authentication is the access control one, in other words to ascertain the privileges of a particular user whose identity has been checked, in order to use a network resource.
- *privacy in communication*: since all network traffic is in clear, sensitive data and messages privacy is a crucial problem that can be solved by encrypting data with suitable secret or public key algorithms.
- *data and messages integrity*: another security problem affecting the Internet is the possibility of counterfeiting network data and messages. The tool used to defend oneself against such attacks is digital signature, to be added to data in order to attest their authenticity. Digital signature, moreover, since it unambiguously identifies the subject that added it to the electronic document, cannot be repudiated.

1.1 Attacks and defence to TCP/IP protocol

The Internet ensures wide connectivity thanks to the use of an open protocol such as the TCP/IP one, which nevertheless does not take into account any support for cryptography and, therefore, is completely insecure as far as the use of applications needing a good degree of secrecy is concerned. In order to solve any kind of network's, and in particular the Internet's, security problem, some network protocols compensating TCP/IP's critical points have been proposed. That is why encrypted variations of transmission protocols were born to protect confidential data transmission.

SSL protocol, for instance, is an interface layer between the transport layer and the application layer in TCP/IP stack, defining a safe communication channel (fig. 1). During handshaking phase, communication entities, client and server, exchange the necessary messages to mutually authenticate themselves by using public key cryptography and establish session keys and cryptography algorithms to be used during the communication session.

A cryptography technique carried out on the application layer can be found in S-HTTP and S/MIME protocols as far as the web and e-mail are concerned. Both protocols redefine and replace HTTP and SMTP protocols. Other protocols are defined above the application layer, such as SET, defined for e-commerce and credit card payments, as well as PGP, used for electronic mail.

1.2 Cryptography

Generally speaking, in order to protect confidential information, we make use of cryptographic systems, i.e. methods to encrypt information, so as to make it unintelligible and therefore useless to anybody who is not interested in it. Encryption is a particular kind of computation and almost all modern cryptosystems base their security on computation difficulty: changes made on data are so complex to make economically prohibitive the decryption cost.

1.2.1 Conventional (or symmetrical) cryptography

In a conventional cryptosystem whoever wants to send a confidential message must provide itself with an algorithm, or a ciphering general procedure G, and

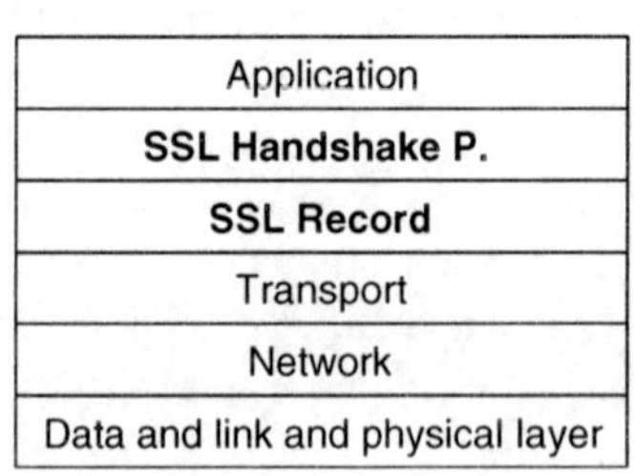

Figure 1: SSL position in TCP/IP stack.

a key K. Being the algorithm G a public one, the security of these systems depends completely on the security of the keys K that enable to effectively encrypt with G_K and decipher with G_K^{-1}, hence the need of resorting to secure channels to distribute keys to the users, with all the inconveniences such system has in terms of effectively possessing, managing and transmitting plurality of secret keys in the hypothesis of confidential communication among various recipients.

1.2.2 Public key (or asymmetric) cryptography

In a public key cryptographic system the parties, instead of using the same key, generate each a couple of different keys:

- a ciphering key k_p to be used in G (public) general procedure;
- a deciphering key k_S, useful to complete $H = G^{-1}$ (public as well) deciphering general procedure.

The user that has generated the two keys k_p and k_S puts therefore the former in a public archive and keeps the latter secret. These two keys are correlated; they specify ciphering G_{k_p} and deciphering H_{k_S} operations that are inverse to each other; but, known the key k_p, it must be impossible to compute k_S.

According to what just said, we can give a more rigorous definition of public key cryptosystem. A public key cryptosystem is a couple of algorithm families $\{G_{k_p}\}$ and $\{H_{k_S}\}$ representing invertible transformations on a finite message space $\{M\}$. In other words,

$$G_{k_p} : \{M\} \rightarrow \{M\}$$

$$H_{k_S} : \{M\} \rightarrow \{M\}$$

so that

a. for all the keys k_p and k_S, effective ciphering G_{k_p} and deciphering H_{k_S} algorithms must be inverse to each other so that

$$H_{k_S}(G_{k_p}(F)) = P;$$

b. for all keys k_p and k_S, both G_{k_p} and H_{k_S} must be easily computable;
c. by publicly revealing k_p and therefore G_{k_p}, the user does not reveal an easy system to compute H_{k_S}. This means that actually only the user can decipher G_{k_p} ciphered messages or effectively compute H_{k_S};
d. if a message is first "deciphered" and later "ciphered", P is still the result, i.e.

$$G_{k_p}(H_{k_S}(P)) = P.$$

Let us examine the above-mentioned four properties. If condition (c) occurs, any crypto-analyst – not being able to derive H_{k_S} from G_{k_p} – in order to trace the clear text back, must take into consideration all possible messages $\{P\}$ and then verify if there exists a particular message P_X whose result is $G_{k_p}(P_X) \equiv C$. In other words, we would face what in cryptanalysis is called a *brute force* attack. The number of clear messages to be examined is, however, so high that it is actually impossible to trace the secret text back through this approach. When conditions (a) and (c) occur, it is sure that G_{k_p} is a unidirectional function. Condition (d) is instead necessary to carry out the so-called "digital signatures". If user A wants to send a message to user B, he "deciphers" it exploiting his own k_S^A, deciphering secret key thus obtaining $H(k_S^A, P) = S$ (signature), and therefore ciphers (in order to ensure secrecy) the latter exploiting the receiver B public key, i.e. k_p^B, obtaining the encrypted text:

$$G(k_p^B, S) = G(k_p^B, H(k_S^A, P)) = C$$

The recipient B – once received C – operates with its own deciphering k_S^B secret key, obtaining:

$$H(k_S^B, G(k_p^B, S)) = S$$

Then it "encrypts" the result with the sender A's public ciphering key k_p^A, and obtains:

$$G(k_p^A, S) = G(k_p^A, H(k_S^A, P)) = P$$

Since A only could have generated a message with that property (it only knows k_S^A), the signature problem is solved. Moreover, receiver B cannot modify P in a different version P′, because in that case it should create the corresponding signature $S' = H(k_S^A, P')$. Message integrity is therefore ensured.

A possible representation of a public key cryptosystem flowchart is shown in fig. 2 [1].

The critical point when planning such a system is to find a couple of general procedures, G and H, for which a couple of inverted keys k_p and k_S, so that k_S cannot be computed from k_p, can be easily detected. A source of such couples is made up of a group of mathematical problems that complexity theory has characterized as non-deterministic and in polynomial time or NP problems.

1.2.3 Hybrid cryptographic systems

Public key algorithms do not actually replace private key ones. In most practical implementations, as a matter of fact, public key cryptography is used to safely spread *session keys*, in other words different keys generated for each new communication session and used just once to exchange messages through symmetric algorithms. This is mainly because public key algorithms are at least 1000 times slower than private key ones. Such systems are called hybrid systems and their use solves a relevant key management problem. Session keys are in fact created when they are needed and destroyed after their use. This drastically reduces the risk of compromising the key.

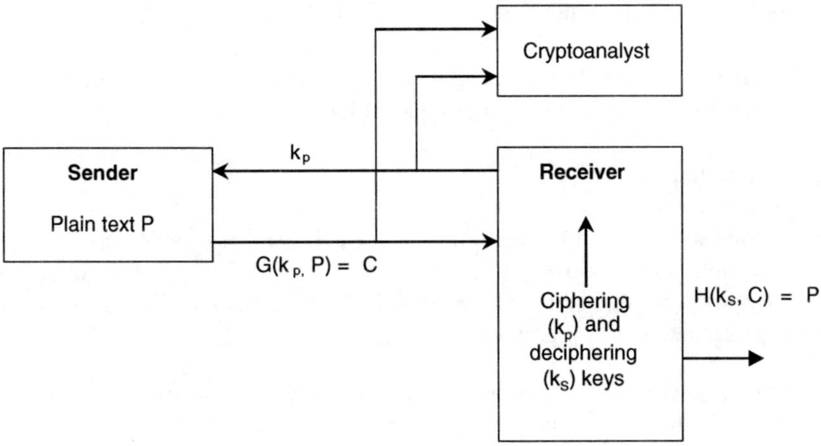

Figure 2: Public key cryptographic system.

1.2.4 Public key algorithms

The main ciphering public key algorithms (that are, among other things, the basis of so-called system for *digital signature*) are based on the computational difficulty in factorizing great positive integers or calculating discrete algorithms. The most important one is the RSA system that is the industrial standard.

RSA system is based on a traditional NP problem, i.e. that of factorizing great positive integers. Such system is based on the fact that, though it is easy, from computational point of view, to find great integers x and y, factorization of product $P = x \times y$ of such numbers is, up to now, *computationally infeasible.*

The algorithm security is not absolute and, on the contrary, the effort to break down a typical RSA key with a brute force attack can be quantified and depends first of all on the length of the public key chosen. Typical RSA public keys are made up, usually, of a precise number of bits: 384, 512, 768, 1024, 2048. As the public key length increases, also the effort needed to factorize it increases. RSA system, actually, is absolutely resistant, thanks to the particular importance of crypto-analytic problem, among all computation problems.

From the performance point of view, stated k equal to the bit length on number n, RSA algorithm has magnitude of computation time proportional to $O(k^2)$ for the ciphering phase, to $O(k^3)$ for the deciphering phase and to $O(k^4)$ for generating keys [2].

1.2.5 Symmetric algorithms

Among the most widespread symmetric encryption algorithms used up to now, we recall the data encryption standard or DES, its successor 3DES and the *Rijndael* algorithm. The latter has been chosen as the new standard symmetric algorithm for cryptographic applications in the place of DES. Rijndael algorithm is a symmetric block cipher supporting keys of 128, 192 or 256 bits. Rijndael's structure shows, above all, a high degree of modular design, which should enable to make the necessary changes to deal with any future attack, in a simpler way than the traditional algorithms planning techniques [3].

1.3 Authentication

Cryptography's aim is the solution of practical problems involving secrecy, authentication and integrity. A cryptographic protocol is a protocol using cryptography, which is used to prevent or detect spying and falsification attempts. There are various kinds of protocols:

- *protocols with arbiter*, in other words, disinterested third parties (that do not have particular bonds with any of the parties involved) and whom everybody trust;
- *protocols with judge*, in other words a disinterested and trusted third party (TTP), as the arbiter, but not directly participating in every protocol: it is summoned just in case of dispute;
- *self-enforcing protocols*, planned to guarantee impartiality and fair play. There is no need of judges or arbiters (if one of the parties tries to cheat, the other one notices it and protocol ends). Unfortunately, they are not always applicable [4].

1.3.1 Authentication protocols

Passwords are still nowadays the most used tool in authentication phase, but they are subject to various threats at security level. If we consider, for instance, login attempt to a system with *brute force* attack, it can be estimated that the password can be guessed with a success probability P equal to:

$$P = \frac{L \times R}{S}$$

where L is the password lifetime, R the attempts frequency and S the password keys space [5]. For such reasons, modern computation systems make use of cryptographic authentication systems. In this case, authentication service enables to check that participants to data communication session are really who they claim to be, and authentication generally occurs during the phase establishing communication.

1.3.2 Authentication based on shared secret key

Given two equal entities intending to communicate on a secure channel, most authentication protocols – called challenge-response protocols – take into account that one of the parties sends a chance number to the other (challenge), which suitably transforms it (so as to authenticate itself) and later returns the result (response) [6].

1.3.3 Authentication through public key cryptography

Authentication through public key cryptography is based on Diffie–Hellman's protocol allowing the exchange of a secret key on an unsecure channel (fig. 3) [7]. The aim of the opening exchange is therefore that of authenticating one another and agreeing on a secret and shared session key. If A and B are the equal entities intending to authenticate themselves, they will proceed as follows:

- A starts by ciphering its identity and a chance number R_A, using B's public key, i.e. it calculates $C(K_P^B, A, R_A)$;
- B returns A a message containing A's number R_A, its own chance number R_B and a session key k_S;
- A deciphers the message using its own private key, finds R_A and thus it is certain of B's identity;
- A accepts the session by returning $C(k_S, R_B)$;
- when B receives $C(k_S, R_B)$ it knows that A has received the message 2 and checked R_A.

The protocol, though, assumes that A and B know each other's public keys: certainly not a trivial matter that refers back to key certification problem.

1.3.4 Authentication systems

Protocols hitherto shown are at the basis of the main authentication systems using cryptography. The essential features of such systems are briefly presented [2]:

- *kerberos* is a key authentication and distribution system developed by MIT; it is based on a Needham–Schroeder variation. It extends authentication, authorization and access system to the whole network involved;
- *NetSP* is a key authentication and distribution system developed by IBM. NetSP's peculiar feature is the use of one-way hash functions to check the integrity of exchanged information;
- *SPX* is the key authentication and distribution system developed by DEC. It is a hybrid system based on the use of both secret key cryptography (DES

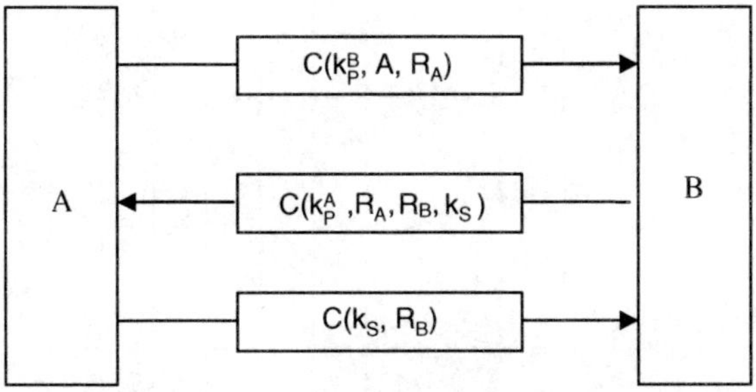

Figure 3: Authentication through public key cryptography.

Table1: Authentication systems comparison.

System	Security services					Cryptographic techniques			Availability	
	Authentication	Privacy	Data integrity	Access control	Non-denial	One-way hash	Secret key cryptography	Public key cryptography	Public Domain	Commercial
Kerberos	X	X	X			X	X		X	X
NetSP	X	X	X			X	X		X	
SPX	X	X	X			X	X	X		X
TESS	X	X	X		X	X	X	X		
SESAME	X	X	X	X		X	X	X	X	X
OSFDCE	X	X	X			X	X	X	X	

algorithm) to guarantee privacy and integrity, and public key cryptography (RSA algorithm) for authentication;

- *TESS* provides a universal security toolbox for access control, authentication, key exchange, privacy, digital signatures as well as for security management verifiable in network distributed environments;
- *SESAME* is a European community research and development project. Authentication model appears as an extension of kerberos functionalities, through the use of public key cryptography algorithms;
- *OSFDCE* is a collection of integrated tools and services to support development and use of Open Software Foundation (OSF) distributed applications.

Table 1 shows a comparison among the above-mentioned authentication systems.

1.4 Digital signatures

From a technical point of view, *digital signature* is a sequence or "string" of data resulting from a ciphering operation on a digital file (e-mail, text document, image, code, etc.), by using the double-key cryptography method. It guarantees data *authenticity* and *integrity*.

1.4.1 Digital signature process

Digital signature process develops through several phases that can be thus briefly summed up (fig. 4):

- *digital fingerprint extraction*: the first thing to do to produce a digital signature is extracting from the original file the so-called "*digital print*", i.e. a string

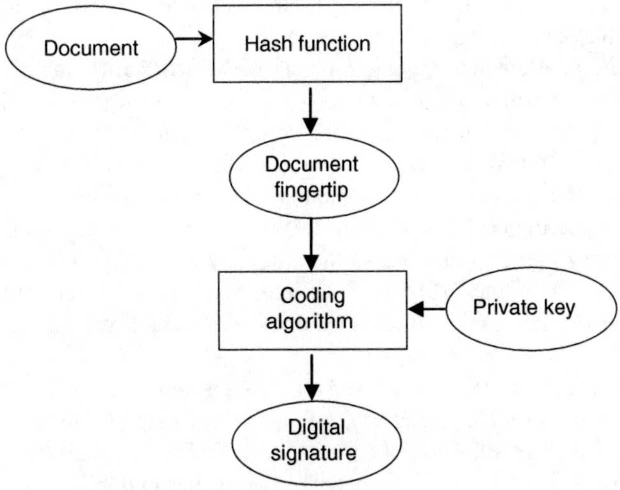

Figure 4: Digital signature generation.

of data univocally summing up their content. The print is generated through particular algorithms, based on the so-called *one-way hash functions* which we will deal with in the next section;

- *private key print ciphering*: the second operation is the print ciphering through the signer's private key the result of which is the actual digital signature, added to the original file. Different algorithms can be used to generate and verify digital signatures. The most common ones are the RSA, ElGamel, Schnorr and DSS algorithms;
- *digital signature sending and checking*: once generated the digital signature, the sender can send the message "in clear" with annexed digital signature and, if needed, the certificate through which trace back the public key's value. If the sender also has a privacy problem in communication, it can cipher everything with the receiver's public key and only the latter will be able to decipher it (with its own private key);
- *temporal marking*: when it is necessary having certainty about the document's time of creation and validation, temporal marking can be used. This consists in generating, by a third trusted party, usually a Certification Authority, of a further digital signature other than the subscriber's one.

1.4.2 Asymmetric cryptographic algorithms for digital signature

RSA system is actually the industrial standard for the generation of digital signatures. The use of such system is simply based on the inverted key roles of privacy ones. Other systems use different signature algorithms than ciphering ones. Among these let us remember ElGamal, Schnorr and DSS algorithms that owe their strength to the discreet logarithm problem complexity [8].

1.4.3 Key certification

Within the systems used for digital signature it is necessary to further outline a so-called public key cryptography infrastructure (*public key infrastructure,* PKI) regulating the relationships among subjects with certification tasks [9]. PKI is based on *hierarchical model*, resorts to the intervention of the so-called trusted third parties that are in a neutral position compared with digital signature users. The *Certification Authority*'s task is, therefore, to guarantee that the public key of a certain subject effectively corresponds to the reference entity. This system is adopted by the international standard X.509 outlining an articulated Certification Authority hierarchy acting as the authority for the community of users; it is adopted in PEM and S/MIME for e-mail and in SSL for channel security. Such mechanism is also used to certify software (Java and ActiveX) components signature.

The existence of more Certification Authorities is moreover taken into account; they are arranged in a tree structure with a CA entrusted by users at its roots. Certificates issued by a CA are checked and certificated in their turn by higher hierarchically ordered CAs. The CA at the top of the hierarchy guarantees the "trust path" of this PKI model.

1.4.4 One-way hash functions

An authentication scheme not requiring the entire message ciphering is based on the idea of a hash function (not invertible as well as collision-free) transforming the message in clear M and calculating from it a string of bits of assigned length (usually 128 or 160 bits) called digital print.

Message synthesis naturally works with public key cryptosystems: the sender first calculates in clear message M's print and later signs it, sending the receiver both the signed print, let us say F, and the text in clear M. As in all ciphering procedures, the longer the print generated by the algorithm, the more security increases.

There are various generation algorithms, e.g. MD5 generating 128-bit *digest*, Secure Hash Algorithm (SHA-1), accepted also by ISO in the norm ISO/IEC CD 10118-3, able to generate 160-bit digest.

2 Mobile agent systems security models

The analysis of mobile agent features leads us to detect two main categories of security problems. The former concerns the execution environment security. The latter includes problems raised by the agent protection. Agents and execution environments must see each other with suspect, as far as a certain degree of mutual trust cannot be proved.

The agent can try to take possession of or destroy protected information, overcrowd or compromise some resources with the only aim of hampering the operations regular development. The execution environment must be able to authenticate the agent and give it a certain trust; in other words, it must ensure the agent the right to access and use a limited and predefined number of resources. The execution environment, therefore, in order to prevent confidential information thefts or damages, must be able to prevent every agent's attempt at violation of access limits.

Execution environments, in their turn, can try to extract precious information carried by agents and influence their behaviour by modifying their data or adding some code. This leads to the need of defining a mobile agent security model.

The distinction between "policies" and "mechanisms" is very important to build a flexible model: policies establish *what* needs to be done; mechanisms, on the contrary, say *how* things can be done. In order to simplify the problem and refer it to a mobile agent environment, we could state that: the security *problem object* in a system is the usable set of resources forming it (be them physical resources or information); the *problem subjects* are the users (be them agents, processes, servers) that can have access to the system and therefore to its resources; the *problem solution* is the adoption of a suitable security policy that, through the use of suitable mechanisms, ensures the protection of the various resources from illegitimate accesses.

Besides physical resources and information, in distributed systems there is a third category of resources to be taken into consideration: communication channels. It must in fact also be ensured the protection of information passing through a channel composed of a network of hardware and software components that are not under direct control of a single authority.

3 Attacks to mobile agent systems security

We will now deal with the security problem as far as a mobile agent system is concerned, and we will describe the various kinds of attacks that an agent or a platform hosting it can undergo [10]. In the next sections, we will give solutions for agents security and we will focus on, though briefly, on platform security.

Attacks to mobile agents security can usually be classified into four categories: information revealing, denial of one or more services, information corruption and interference. We will use some agent system's components to detect attacks, identifying the possible source and target of the attack. It is important to notice how many of the problems dealt with can be found in client–server systems and have been already treated, in other forms, earlier. Mobile agent paradigm has, nevertheless, the unique characteristic that, contrary to traditional security problems where the application and the system had the same owner, now the agent's and the system owners can be different.

In order to describe an agent system, we will use the simplest model, in other words the one formed by the agent and the platform hosting it. Every agent needs to carry with it some data and to exchange them with other agents. The agent platform provides the environment in which the agent works, and the latter, thanks to its mobility, can move from a platform to the other. A server hosting some agents must give them the possibility to interact. All those aspects have security implications.

We can distinguish four categories of attacks, according to the aggressor's and the victim's identity:

- an agent attacking a platform;
- a platform attacking an agent;
- an agent attacking another agent on a server;
- other kinds of attacks to an agent system.

3.1 An agent against a platform

An agent entering the platform has two main lines of attack. With the former an unauthorized access to the server's information is carried out; with the latter an authorized access is used to damage the platform. Unauthorized access could be carried out simply thanks to the lack of control mechanisms, or because of a certain weakness in the server's identification and authentication systems; this enables an agent to enter the platform pretending that it is another entity. Once the access obtained, the server's information can be read or modified. Together with confidential data, such information could include some platform's code instructions and, according to the access level, the agent could be able to completely stop the platform. An agent could moreover deny some services or resources to other agents, if the rules for their use have not been established or well recognized.

3.2 A server against an agent

When it gives hospitality to an agent, a platform can easily isolate it and "capture" it, taking its information, damaging or modifying its state or code, denying service requests, or simply re-initializing it or ending it completely. Stealing electronic money from the agent is a typical example. An agent is very susceptible to a platform and can be harmed by the server through false responses to information or services demands, making changes to external communications or making it waiting unceasingly. Extreme cases include complete analysis and change of the agent's inner structure, so as to introduce more deceitful changes. The platform's modification of the agent is a particularly insidious form of attack, since in such way the agent's behaviour as well as its computation correctness can be completely changed.

3.3 An agent against another agent

An agent can attack another agent by using different techniques. They include transaction falsification actions, conversation interception or interferences in the agent's activities. An agent that has been attacked, for instance, can answer in a wrong way to direct requests, or deny what is needed to correctly carry out a transaction. An agent can obtain information disguising itself as broker, or using the platform hosting it to intercept messages in transit. When platform is weak or does not make use of all necessary control mechanisms, an agent can have access to another agent's code and state, modifying it, or interfering with it through its public methods. Furthermore, though there could be some control mechanisms in the server, an agent could repeatedly send messages to other agents, so as to hamper their communication.

3.4 Other entities against an agent system

Though neither the agents nor the platform can have malicious behaviours, other entities, inside or outside the framework, could achieve some actions capable of harming or subvert the agent system. The most common methods are those aiming at the interception of communications among agents or inside or outside the platform (agents that communicate from different platforms). For example, at the level of agent-to-agent or server-to-server protocol, an entity could intercept messages in transit from or to an agent or a platform, in order to obtain information. Such messages could furthermore be modified, replaced or simply sent again later, with the aim of disturbing synchronization or the framework integrity.

4 Protocols and techniques for mobile agents security

An attack is a violation of the programmer or the agent owner expectations caused by one or more intentional attackers. Excluding all cases in which violation is caused by exceptions such as faults or technical defects, various kinds of attacks can be classified (fig. 5) [11]. In *reading attack*, aggressors reading or copying private data do not leave any detectable trace, in contrast with the *writing attack* that implies the aggressor's modifying the agent's code. In the first case, the reading of private data carried by the agent can jeopardize privacy (violated keys) or cause economic damages (electronic cash desk). In the case of reading of the agent's code – through the knowledge of its execution strategy – the aggressor can monitor the victim's transactions in order to achieve other ones to its own advantage. Finally, from code and data knowledge, the aggressor can simply infer – from the analysis of the execution steps – further information on the agent's state.

4.1 Solo and team attack

In contrast with solo attack, many aggressors must collaborate on team attack. This kind of attack may also include both reading and writing attacks. For example, many hosts can collaborate to follow an agent's route and obtain information on it or its sender. Another example concerns the case of an aggressor that has modified mobile agent's code or data, while the other aggressor (accomplice) ignores it. This category also includes denying service attacks.

4.2 Unintentional attack

In contrast with intentional or effective attacks, unintentional or superficial attacks are the ones that have not been planned. The aggressor can modify or delete some parts of the agent without knowing what actually has been changed or deleted. Furthermore, the aggressor does not know the effects of its attack. In extreme cases, aggressors can totally destroy the agent or make their execution denied. Total destruction of the agent can be classified as a total attack. In this case, the agent cannot achieve its task.

4.3 Current protection schemes

In order to fight attacks against agents many protection schemes have been proposed, most of which are effective against many attack classes. Of course they cannot detect who attacks, since reading attacks do not leave any trace. In particular, if the protocol is based on the amount of confidence placed in the next host of the agent's route, in order to control the mobile agent's state along the way, such protocol is ineffective in the case of team attack, if two consecutive hosts collaborate [12].

4.3.1 Passive protection

This protection scheme protects the agents by using organizational or architectural solutions. Such approaches avoid attacks by taking into consideration the hypothesis that the execution environment is protected, otherwise they are vulnerable to most attacks.

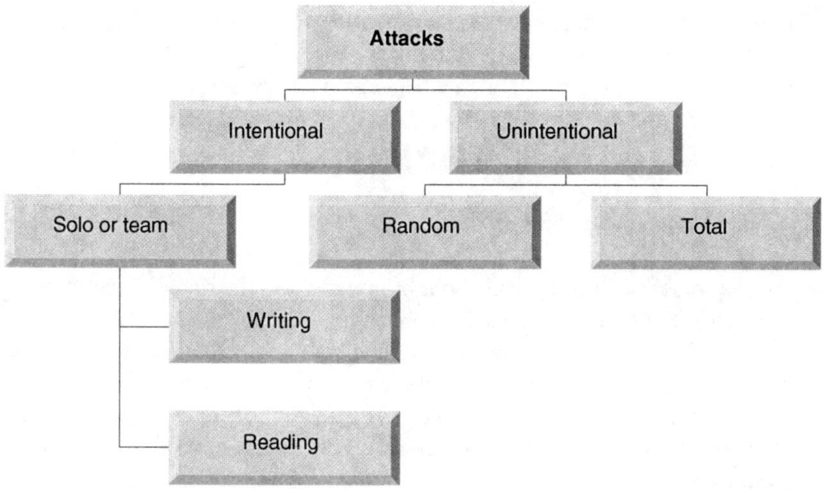

Figure 5: Scheme of possible attacks of a host against an MA.

4.3.2 Active protection hardware

Safe hardware devices for agent protection are used in this case. Using trusted specialized hardware, the same host executing the mobile agent cannot have access to its code and data. It reduces solo attack risks but cannot be effective for team attacks. For example, a malicious host directs the agent towards another cooperating malicious host without the owner discovering it. In such case, to fight this kind of attacks, other approaches must be taken into consideration.

4.3.3 Active protection software

This class implies many protection schemes providing various protection levels. Some schemes take into account the agent's encrypted data, so that only the recipient can decode them using a public key cryptographic system. Mobile agents also make use of cryptographic techniques to encode received data so that potentially malicious hosts cannot violate data. Such approaches though cannot fight against writing attacks.

4.4 Protection schemes under development

Under development protection schemes aim at improving several features. Since not all applications need the same set of countermeasures, they can be applied according to the threat profile and the application security goals. For unintentional attacks, which are difficult to consider in advance, they being quite similar to chance errors, some fault tolerance approaches can be used to detect and correct mistakes as well as to increase the agents' survival rate. Finally, in order to totally protect the mobile agent from reading attacks, an excellent solution would be the adoption of a hardware-dependent model. This implies, however, high production costs and scarce scalability. On the contrary, in order to fight against reading attacks, software solution offers many ideas for further research.

5 Agent protection protocols

5.1 Agent's integrity protection

One of the most sensitive issues on mobile agents is the one concerning data integrity. Several protocols have been studied for both prevention and *a posteriori* identification of possible modifications in the agent's data. Data modification prevention is carried out through special hardware or software techniques protecting the agent for a certain time. Here, however, we will deal with some solutions for *a posteriori* recognition of possible modifications of the agent's integrity.

The first one makes use of a TTP, i.e. a server interposing itself between the agent and the server hosting it, verifying each time that integrity is preserved. The second one is based on a distributed approach, without the presence of a TTP, where data security is ensured by a chain structure among the various servers hosting the agent. The third one is a combination of the first two. All solutions are based on the following assumptions [13]:

- it is realistically assumed that only a prefixed percentage of sites visited by an agent can be malicious;
- there is a public key structure: every entity is associated a public key through authority certification. Public keys and related certifications can be looked at by everybody;
- an agent is composed of three parts: code and initialization data (CID), application data (AD) and protocol data (PD). CID is the unchangeable part of the agent, signed with its own private key in order to ensure integrity and impossibility of repudiation. AD contains the data gathered by the agent, possibly encrypted with its public key if secrecy is required. AD's integrity can be *a posteriori* verified through insertion in PD part;
- integrity control state exploits standard cryptographic primitives, such as message integrity code (MIC), which is a particular one-way hash function. In order to protect the agent's code, the sender computes and signs an MIC attached to PD part and registered by the sender itself.

5.2 TTP solution

Let us suppose we are in the presence of a TTP. Let us distinguish the sender site, the TTP and unreliable remote sites (S_i, $i = 1 \ldots n$) (fig. 6). In this solution, every agent being executed in unreliable sites can only insert data in AD part. Before moving to one of the sites, the agent must visit the TTP performing cryptographic

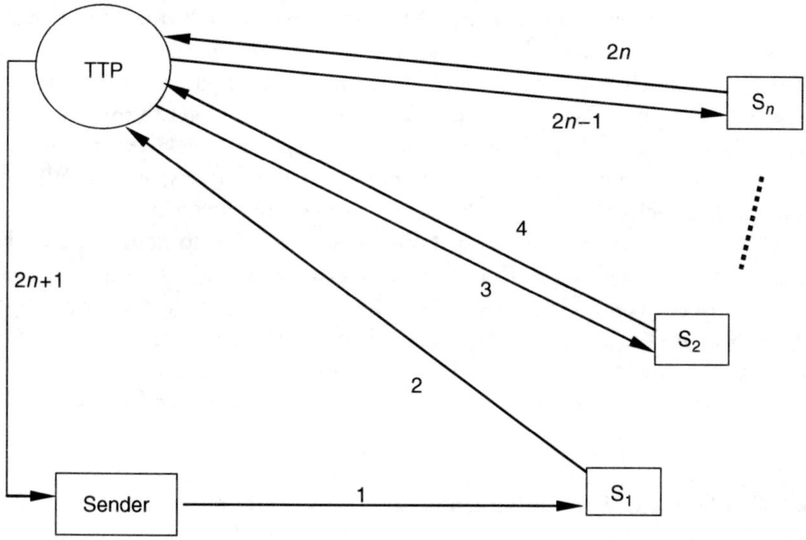

Figure 6: TTP protocol.

operations to ensure new data integrity. In particular, TTP creates an MIC of the previous data gathered by the agent; MIC is put in the agent's PD part and is, moreover, saved by TTP itself.

It is easy to notice that the presence of a TTP ensures data integrity. Data gathered by the agent, in fact, cannot be modified or deleted because every cryptographic function is delegated to TTP. Two malicious hosts collusion, besides, has no effect because they can just modify the AD part of the agent and such modification can be easily detected. Finally, data gathered during the agent migration are available even in the presence of a protocol failure and, therefore, to a break in the visits due to malicious interferences.

It must be noted that the agent is not obliged to visit always the same TTP, because the signed MIC, being incorporated in itself, can ask to carry on the control service to other TTPs. Protocol nevertheless expects the agent to migrate to TTP after every visit to an unsafe site, which causes a great overhead. TTP protocol scalability, moreover, is undoubtedly limited by the effective availability of reliable sites. Generally speaking, TTP presence can be taken into consideration when agents move across sites having a certain mutual reliability (e.g. sites belonging to the same organization), unrealistic characteristic when we talk of the Internet global environment.

5.3 Multiple jumps protocol for agent integrity (MH)

MH protocol is a distributed solution ensuring the integrity of the agent's state without the need of any TTP. The agent gathers partial data from every site and attaches the previous data in its own AD part. Each site, in its turn, must produce a brief *proof* document of the agent's computation, which is saved in the PD part of the agent itself. The main idea is to connect, through cryptography, every proof with the one created by the previous site during the agent's migration. This establishes a chain relation among proof documents and makes it impossible to modify an intermediate proof without modifying the whole sequence. When the agent comes back to the sender, the latter can check the integrity of the encrypted proof chain and find possible violations. It is interesting to note that the chain relation among ciphered proofs is such that a malicious site S_i is not able to change or delete intermediate data D_k with $1 < k < i$, without being discovered by the sender. Let us note, however, that, as any other distributed protocol, MH does not prevent the malicious site from totally delete/replace all data in a chain through falsification. Such protocol moreover introduces overhead due to data signature operations.

5.4 Combined TTP and MH protocol

TTP solution has more fault tolerance than MH, because there are no relations among successive visits to sites: when network problems or malicious interferences occur, intermediate results can be available to applications. Anyway, TTP requires the presence of reliable sites within the agent's field of action, and those are onerous terms. When they are not available, the agent's integrity can be saved by MH protocol.

When migrations are many and sites are connected in different unrelated domains, neither TTP approach nor MH one offers a good solution in terms of efficiency, strength and scalability. In this case an approach integrating TTP with MH can solve both protocols' problems.

In a combined approach, the agent's itinerary is partitioned in small partial paths. Within a path, the agent's integrity is secured by MH protocol, and the agent is required to migrate to TTP only after the visit to the partial path last site. TTP verifies the integrity of gathered data and computes a new protected state of the agent's. It moreover generates a new secret value to be used in the next route and encrypts it with the next site's public key. This avoids the considerable reduction of performances that instead occurs applying a pure TTP protocol, reducing the visits to TTP itself. The use of TTP protocol increases the strength of MH one: the chain of data is partitioned through intermediate controls (the various visits to TTP) recording the application partial state. Combined solution increases the application flexibility ensuring the agent's state integrity for a wide range of applicative scenarios.

5.5 OKGS (One time key generator system)

A method similar to MH protocol, which can therefore be an alternative, is the OKGS (one-time key generator system) [14]. Here, when a server receives an agent, it checks its digital signature. Later it generates a new secret key, an agent-key, based on a "source key". Source key is generated by the value C_{k-1}, that has been provided by the previous server. The server encrypts the data generated through the agent-key, which are saved in the agent's work area. This means that every agent encrypts its data. At the same time the server generates a new value C_k, which will be used by the next server. Then, the agent migrates on a new server. Thus, an interrelation between consecutive keys can be created. The data encrypted in the agent's work area are deciphered by a sequence of agent keys. When the mobile agent goes back to the sender, every key can be regenerated since each key creation is based on a one-way hash function as well as on every secret piece of information in the server, which is encrypted with the sender's public key. Thus, the sender can generate every key used in every server from the source value that had been given to it.

In OKGS, digital signature and timestamp play an important role for code and data integrity preservation. Digital signature in this system is used in two ways: the agent's code is signed by the sender server to ensure integrity; on the contrary, the agent's data, during its journey, are signed by every server to guarantee the identity of hosting sites. Moreover, the server marks such data with a timestamp by attaching current date and time value: thus, "replay" attacks, i.e. attacks exploiting already used signatures, are avoided.

OKGS protocol is based on consecutive agent-key interrelation, which makes it very similar to MH protocol. OKGS protocol makes, however, use of data timestamp as well, which avoids possible attacks exploiting already used keys. Weak points of OKGS system are connected to the security of hash function and cryptographic system, which can be broken by cryptanalysis attacks based on key search.

6 Environmental key system

One of the defects in traditional cryptographic systems is the static nature of secret keys: they do not depend on time, space or operational conditions. On the contrary, information secrecy is strictly linked to these conditions as far as mobile agents are concerned. Suppose, for example, that a customer wants to buy a product or a service that must be revealed exclusively to the site that makes it available, while all other hosts receiving the agent during his search are not allowed to know anything about it. Actually the agent itself does not know its goal, but it must be created so that, as soon as it finds the right server, i.e. the one having the product/service looked for, it is activated and can make the purchase. All that can be achieved using the notion of environmental key generation [15]. The agent, using such data- and code-encrypted keys, would remain inactive until it finds itself certain environmental conditions.

The approaches for the safe creation of cryptographic keys from environmental observations are basically three. The first one directly involves environment manipulation in a cryptographic way. In second one, the trusted server is used. The third one makes use of dissimulation of a certain environment nature of values carried out through the use of a one-way hash function.

6.1 Agents "in the dark"

An agent has typically got a ciphertext (a set of data, a series of instructions, etc.) and a method for the search through the environment, of necessary data to generate decryption keys. When environmental information is located, the key is generated and the text is decrypted. Without environmental inputs the agent cannot decrypt its own message (the agent actually is in the dark about its goal), and is therefore resistant to an analysis determining the agent's function.

Let us assume, therefore, in our formalization, the following hypotheses: N be an integer corresponding to an environmental observation; H a one-way hash function; M the result of H applied to N, necessary for activation; $\oplus$ the operator *or exclusive*; the "comma" the concatenation operator; R a particular moment; & the *and* operator; and K a key. The value M is carried by the agent. Hash function can be used to make tests and build one-way keys, so that an agent exam does not reveal the required environmental information. Here are some possible constructions:

```
If H(N) = M then K = N
If H(H(N)) = M then K = H(N)
If H(N_i) = M_i then K = (N_1, ...,N_i)
If H(N) = M then K = H(R_1,N) ⊕ R_2
```

According to the different settings of the application, a different construction can be used: the important thing is that it is impossible for each construction to trace back the key K through the knowledge of M.

6.2 Basic construction

Through the use of analysed constructions first we realize a basic version of our environmental key system. As we have seen, N, which is directly linked to an environmental factor, is subject of the hash H function. The agent, therefore, in its journey does nothing else but analysing every time the particular environmental conditions that become subject of H and are compared to M: if the value coincides, it means that the environmental condition is the right one and the key K can be generated. Let us notice that the environmental conditions used to generate keys can be very different ones: a particular image contained in a web page, a file with a particular name contained in a server's file system, a particular message among the various present in a newsgroup visited by the agent.

6.3 Temporal construction

This construction enables key generation according to a certain time value. Protocol needs the presence of a TTP that does not need to know both parties involved, nor the nature of data for which keys have been created. Protocols of such kind have three different phases:

- programmer–server interaction, where the programmer obtains an encryption key from the server;
- programmer–agent interaction, where the programmer provides the agent with the encrypted message, some required (but not sufficient) decryption data, as well as information concerning where to go to decode the message;
- agent–server interaction, where the agent gathers the data necessary to create the key and decipher the message.

Two among the most used time-based protocols are the *forward-time* one enabling key generation only after a given time, and *backward-time* only before the given time. These protocols can be combined to generate the key only during a certain period of time. The main defect of such techniques is that the server could be in collusion with an attacker to analyse the agent. That defect can be easily eliminated using the methods discussed below.

7 Resistance to attacks

In a distributed computation system or an intrusion detection system, cooperating or communicating mobile agents find themselves in execution on different platforms. In order to make agents resistant to attacks, some particular strategies need to be adopted. One methodology meant to make mobile agents resistant to attacks is divided into five stages [16]:

- *agent location randomization*: First all agents location is randomized, so that the attacker finds it difficult to detect the agents having critical functionalities;
- *central service directory removal*: Later, the directory containing the list of available platforms, of active agents, as well as of malicious platforms is removed, so that there is not a single statistically located point that can cause the agent-based system fall. Information contained in service centralized directory is included in the agents themselves that, thus, keep trace of the location of other agents they need to communicate with;
- *eluding aggressors*: Then, the agents are informed of possible attacks from other agents as well as of their source. Agents surviving an attack avoid possible aggressors locations and report the other agents on the most dangerous sites in the network;
- *killed agents recovery*: Backup agents replace the agents destroyed during the attack using all their information or partial state. The destroyed agent substitute is chosen among the many available backup agents;
- *cut off communication lines recovery*: Later that substitute looks for other agents in the network and tries to re-establish communication. Through that technique we allow distributed applications to be attack resistant.

7.1 Agents location randomization

Agents can ask their mobile agent platform the other mobile agent platforms list and choose one at random. Alternatively, every agent can carry with it a list of mobile agent platforms. Thus, there are agents that incessantly and randomly move through the network, so that an attacker has difficulties in exactly locating critical agents locations. The attacker can inspect a network and find all mobile agent platforms, but it will be impossible for him to detect which agents reside in each platform. A possible way against attacks to the system would be the continuous monitoring of mobile agent platforms communications, but to deal with that, it has been assumed that mobile agent platforms encrypt all communication traffic. Thus, an aggressor can just estimate how many agents are there on the target platform and determine which platforms contain agents willing to communicate with agents present in the target platform. Thus, if an aggressor has got enough sniffing points, it can be deceived by mobile agent platforms if they send false and random transmissions to other mobile agent platforms.

7.2 Removal of centralized service directory

An aggressor who is not able to locate critical agents locations can, on the contrary, try to disable the entire mobile agent system. That can be done by turning off all network mobile agent platforms or by locating the single weak point which the agents depend on: PKIs and centralized service directories. Dependence on a PKI can be removed by ensuring that every mobile agent platform locally keeps worksheets copies.

On the contrary, a mobile agent centralized directory is difficult to remove. Even though mobile agent service directory tables can be locally saved on every agent platform, when an agent moves it should communicate to all platforms its movements, which is ineffective. The solution is to provide every agent with a team of "friends". A "friendship" relation is mutual and defines a group of agents incessantly notifying each other their movements.

Although centralized directory removal is thus still possible, that has the side effect of producing problems, most of which lead to the agent's deadlock. Using a concurrent negotiation algorithm can solve that problem.

7.3 Eluding aggressors

An aggressor can kill some agents if it takes possession of a mobile agent platform. Agents surviving the attack bypass the network suspected areas that host the aggressor and inform the other agents of dangerous areas. The choice of network areas to be avoided depends on how the aggressor did choose its target and whence it did launch its attack.

Agents must take into consideration three cases of attack: *random choice, knowledge gained through mobile agent platform control* and *knowledge gained through network traffic interception*. In the first case, it is assumed that the aggressor has destroyed a platform through random choice of a host. In this case, mobile agents do not increase their security if they move. In both the second and third cases, it is assumed that an aggressor has destroyed the mobile agent platform knowing that important agents were hosted on that platform. An aggressor can obtain such information only by violating a server, or through intelligent analysis of encrypted network traffic schemes. There are no other methods through which an aggressor can discover which agents reside in a platform.

7.3.1 Knowledge gained by controlling a mobile agent platform

If the aggressor controls a platform that has been violated, he then knows which agents are communicating with the ones that are on his platform, and knows which agents have recently left his platform for the next target one. Moreover, he knows if one of the agents on his platform is communicating with another platform agent. That is the only way to detect the agents located on another mobile agent platform, since it has been assumed that an agent platform cannot ask another platform its list of agents.

An aggressor controlling a violated mobile agent platform – let us remember – cannot launch malicious agents to find information, since mobile agent platforms can only accept digitally signed agents from a security authority.

7.3.2 Knowledge gained through network traffic interception

If an aggressor intercepts network traffic, it can be able to estimate the number of agents on a particular host, by looking at the number of connections and the amount of traffic to and from a given platform. Let us assume that the traffic between two mobile agents in different LANs can only be intercepted by an aggressor located in one of the two LANs. The agents that have survived an attack are not able to recognize the method used by the attacker to choose target mobile agent platforms; therefore, they avoid LANs that could host the aggressor. In particular: movement does not increase the agent's security.

An agent must avoid all LANs hosting mobile agent platforms that have recently communicated with target mobile agent, as well as LANs hosting target mobile agent platforms.

7.4 Recovery of killed agents

In the case of an aggressor killing an agent, the agent must be recovered. That is achieved by keeping, for each agent, clone backup agents in constant communication with the original one. Such communication connections update backup copy and inform it if the original is killed. When an agent dies, backup copies negotiate for the original's place. The backup copy chosen will take on that role, while the remaining ones die.

While restoring a killed agent is simple, there are some complications to manage. If an aggressor is on the same LAN as the original one, when the aggressor kills it, it will witness the communication exchange between backup copies. Therefore, an aggressor can kill an agent and, simultaneously, all its backup copies.

A possible solution is that of creating more backup copies than those the aggressor is thought to be able to kill. However, it is a very expensive/costly solution because there must be constant connections between backup copies and the source one. The adopted solution considers that every backup copy and the source one reside on different LANs.

The source one creates multiple backup copies, in the hope that the aggressor is not simultaneously present in every LAN containing a backup copy. The various backup copies created thus increase the source safety and, in the meanwhile, the network computational load.

7.5 Restoring of cut-off communication lines

When a backup agent takes on the source's place, it issues backup copies of itself in order to protect itself, but, more important, it must contact again the other agents. There are two solutions to this problem.

The first one considers that when the agents leave a host, they register a temporary "trademark" on the starting host, indicating the target host. A restored agent can control its friends' last location known and try to find these indications. It should be able to pass from one host to the other until it finds its friends' current location. That provides the restored agents with a quick method to find their friends, though it is not unfailing. If an attacker kills two friends, then the two restored agents cannot use this technique to find each other.

A slower but more efficient method is that of remaining stationary and verifying the presence of friends in mobile agent platforms, until they find each other. If the restored agent finds a friend of his, then, through the latter, it can let know its position. Such system is slow and inefficient, but enables the agents to find all their friends.

8 Safe agent transfer

One of agent system critical aspects concerns the need for a safe transfer of the agent during its migration process through the network. That process can be summed up as follows: the mobile agent decides to change location and communicates it to the sentinel agent running the site where the agent is currently located. Local sentinel re-forwards the request to remote sentinel that takes a decision according to the credentials presented by the mobile agent itself. If request is granted, mobile agent puts itself in inactivity state. The source site sentinel packages the agent's code, state, data and information necessary to rebuild it in the other site. The agent is ended, while its copy is sent to the target site sentinel that, after having verified its authenticity, re-instantiates the agent.

The aim is to define a protocol making this process as much safe as possible, for all parties involved [17].

Mobile agent will have to mind not to be sent to the wrong address or discover that the destination site is a hostile environment where there is the concrete possibility of being "inspected" and "robbed" of information by unauthorized parties. The sentinel agent, on the other hand, must also face the problem of protecting itself from ill-disposed or badly programmed mobile agents reaching the site. What follows is the schematization of a protocol planned to provide more protection to migrating mobile agents.

Parties involved in the protocol:

I:	source site sentinel
R:	target site sentinel
Nrandom:	random number
KI private:	source site sentinel private key
KI public:	source site sentinel public key
KR private:	target site sentinel private key
KR public:	target site sentinel public key
MA:	mobile agent first residing in I's site
CA:	certificating authority
C:	ciphertext
P:	plaintext
E(p,K):	plaintext p encrypted with key K
D(c,K):	ciphertext c decrypted with key K
X trust y:	x trusts what y says

Initial conditions:

- MA gives credit to I;
- I, R and MA give credit to CA;
- all parties give no credit to nobody else.

Steps considered by the protocol for the agent's authentication, transmission and instantiation are:

Authentication and transmission:

- I asks KR public to CA;
- I gives a copy of KR public to MA;
- I generates C1 = E(Nrandom_1, KR public) and sends it to R;
- R asks KI public to CA;
- R recalls Nrandom_1 = D(C1, KR private);
- R generates C2 = E((Nrandom_1, Nrandom_2), KI public);
- I recalls Nrandom_1, Nrandom_2 = D(C2, KI private);
- I verifies Nrandom_1;
- MA rejects KMA private;
- I generates C3 = E((MA, Nrandom_2), KR public);
- I keeps a copy of MA;
- R recalls MA, Nrandom_2 = D(C3, KR private);

Instantiation:

- R instantiates MA;
- MA doubts R with C4 = E(Nrandom$_2$, KR public) Communication and location model for backup agents;
- MA generates a new private KMA;
- R sends I the message of "completed migration";
- I ends the couple MA.

In order to increase security, a chronology agent should be defined. The sentinel agent receiving unknown mobile agents would thus have the possibility of investigating their past behaviour before their instantiation. Such "records" should be generated for each agent in the site; the "detective" sentinel should ask information to the other sentinels.

9 Safety in mobile agent platforms

9.1 Agent against the platform

The concept on which an agent system is based is that a platform accepts and executes developed code everywhere. It is therefore likely that a malicious agent penetrates inside the platform and attacks it. The agent, in fact, taking advantage of the lack of suitable mechanisms for the control of accesses and identification, could disguise as trusted to the platform and obtain the right to access there using it illicitly. Thus the agent can have access to the platform's information, including part of the code, and, according to the range of its penetration, even destroy it. Even without gaining access to resources, moreover, an agent can deny the services of the platform where it has installed itself to other agents, exhausting computational resources.

9.2 Protecting the platform

Countermeasures to avoid attacks to platform include a series of techniques, procedures or other means that limit the vulnerability of the platform that, without suitable mechanisms, is open to every kind of attack attempt. Assumed that the basic principles on which security is based consider that an agent trusts the platform that generated it, whence execution starts, and which is implemented safely; there are usually various techniques to protect the execution environment [18], which we hereafter describe.

9.2.1 Sandboxing

A way of protecting an execution environment against a malicious agent is to narrow the environment in terms of mobile code access rights and privileges. Actually, code is executed in a sort of "sand box", in a distinct domain, where very few things can be damaged. A single identifier associated to each domain checks access to memory and other resources. This approach has been adopted by [20] for Java applets distribution: they are executed in a Java virtual machine (JVM) that is checked by a security policy. Any browser (Netscape Navigator, Microsoft Internet Explorer, etc.) JVM is set up in a way that it does not allow access to file system, nor establishing TCP connections but through the site whence they have been downloaded.

Those are just some of all possible security policies. The problem is that the creation of a "sand box" means setting some limitations to the code, restrictions that, to some particular applications, can be too strict.

9.2.2 Safe code interpretation

The main idea of this security policy is that commands that could be harmful to the platform be made safe or refused by an agent. A dangerous command, for example, is that which considers the execution of a common string of data as a program's fragment.

The most commonly known language security interpreter is probably Safe TCL, used in the first development of the Agent TCL system. It is based on the concept of "padded cell", referring just to such access isolation and control technique. After this first interpretation, a second "safe" interpreter examines the code, before being executed by TCL interpreter, and points out possible harmful commands to platform.

Various safe interpreters can be implemented, so as to create various kinds of approaches. The creation of interpreters based on this policy needs, however, a great ability to avoid too limiting or excessively protective execution environments.

9.2.3 Signed code

A fundamental technique to protect the code is to sign it with a digital signature. Digital signature is a means ensuring the confirmation of the code's authenticity, its origin and its integrity. Usually who signs the code is the agent's creator or user; hence, digital signature is considered as an indication of the authority under which the agent operates.

Since it is not always possible to determine if a code is safe or not, one tries at least to understand whether it comes from the source that is supposed to be its author. The signature, though, is not a certification of the code's execution without faults. Digital signature involves asymmetric keys cryptography and therefore benefits from the structures that make public keys available.

9.2.4 State evaluation

The aim of state estimate is to ensure that an agent has not undergone any kind of violation. The idea is based on the fact that harmful alterations to the agent's state can be predicted and countermeasures prepared before using the agent. Evaluation functions are used to determine what privileges can be granted to an agent on the basis, for example, of unchanged states. Both the agent author and owner produce evaluation functions that are an integral part of the agent's code. When they digitally sign the code, these functions are protected against possible changes. The platform uses evaluation functions to verify the state of an agent as soon as it comes and to decide thus what privileges grant it.

9.2.5 Proof carrying code

This technique implies that the code's producer formally proves that the code he wrote is safe, i.e. it conforms to the security features previously agreed on with the code's user and, therefore, it can safely be installed and executed. The code, as well as its safety proof, is sent to the customer who must verify that the code is effectively safe. Safety statement is directly generated by the code to ensure that the proof sent matches the code that is being executed.

The proof is structured in such a way that it is simple to verify it without any cryptographic technique or external support. This technique can be considered as a valid alternative to "sand boxing". Nevertheless, there are several problems to be solved before considering this approach a really feasible one. Some kind of security policy formalism must in fact be established, as well as automatic support for proof generation and a technique to limit the several proofs that can occur. This technique, moreover, is quite dependent on the customer's hardware and execution environment, which limits its portability.

9.2.6 Path histories

The idea is to keep a record of the agent's previously visited platforms, so that the new platform to which the agent arrived can decide how to execute it and which limits put to resources. Calculating a path history means that every platform must add a datum in the agent's entry indicating its own and the next platform's identity. In order to avoid violations, the signature of a new entry must also include the previous entry in message calculation.

Such technique is strictly dependent on the platform's ability to correctly decide whether trusting or not the previous platform. An inconvenience is due to the fact that verification is costly as far as path history increases.

9.3 A case study: aglets

The *aglets* system is a project developed by Tokyo IBM research laboratory in 1996. The system defines a series of abstractions and behaviours to use mobile agent technology in geographical networks such as the Internet. The term "*aglets*" is derived from the fusion of the words *agent* and *applet*, since API make use of the applets' concepts and terminology. The system is completely written in Java and supports only mobility of the weak kind: after every movement the agent must start execution from a prefixed point, though keeping the values of previous interactions. They can move from a certain host context to another context using the network [19].

Two active objects can be defined:

- the *aglets* themselves, autonomous objects that are assigned, every time they reach a machine, a private execution thread, and that are able to answer messages coming from the outside (also from other aglets);
- the *Context*, indicating the environment where aglets live. The context is a stationary object able to keep aglets in execution and enclose them in a well-defined machine portion, avoiding unwelcome mobile agents intrusions that could infect the machine or take information without the user knowing it.

Aglet's architecture is based on:

- a library of Java classes forming the necessary support for agents development, which the programmer is provided with, written according to applets model and providing as well classes and execution environment for application executed inside the browser;
- a server (*aglet server*) enabling the agents movements between different machines and providing the context;
- a suitably developed protocol called ATP (*aglet transfer protocol*), optimized and dedicated to agents safe transfer;
- a stand-alone one or integrated in a web page user interface for aglet server, through which it is possible to check the state of the server and the agents life cycle.

The aglets life cycle is basically made up of three operations that include birth, life and death. Communication among agents is carried out by two mechanisms:

- *message passing*, the only direct communication technique, synchronous or asynchronous one, taking into consideration the communication partners knowledge without any location hypothesis. In order to simplify its use, some higher level protocols are provided such as *master–slave, messenger–receiver, notifier–notification*;
- *Whiteboard*, a common area in every place where any agent can leave a message, whoever can have reading access and read left messages to find out if it is one of the recipients. This kind of communication is anonymous and asynchronous with location hypothesis. In an agent environment it is a fundamental tool for the interaction and knowledge of unrelated agents.

9.3.1 Security architecture and model

A security architecture must be based on a trusted code imposing the essential parts of security policy, that is to say:

- protect aglets transfers and their communications;
- suitably check accesses to resources and verify agents' executions;
- check that group of aglets do not interfere with other ones;
- check that aglets do not interfere with hosting platform;
- allow the use of different encryption algorithms;
- keep a low amount of encrypted information;
- be compatible with other standard security structures.

As far as security is concerned, aglets uses an authentication protocol that verifies the agent owner's identity and therefore its classification as trusted or untrusted, thus allowing limited access to resources. Migration mechanism moreover verifies, through the use of a message digest, whether some damages have occurred during the transfer, such as a virus introduction or the agents' code manipulation.

Aglets security model describes where and how security policies must be imposed. They are defined in terms of rules by an administrative authority and specify the conditions under which aglets can access the various objects, the authentication of users and principals, as well as which actions can be carried out by an authenticated entity. The *principal* is an entity whose identity (formed by a name and other attributes) can be authenticated by the system where it asks to enter.

Among the *principals* we can find (fig. 7):

- *aglets*: every agent has a single identifier, independent from the context in which it is executed, that is given it at the moment of creation. The agent owner, on the contrary, takes into account the aglet's security, possible violations it can undergo and the possible unreliability of data returned after the execution. For this purpose, it defines some security preferences, i.e. a series of rules establishing who can have access to or interact with the agent during its itinerary.
- *contests*: the contest matches the agent's execution platform. It is usually identified by the host's URL as well as by another element, if there is more than a context in the host. The context master is responsible for its security; he defines some security policies to protect local resources against aglets, but must also ensure that aglets do not interfere with other ones except if that is allowed by the agent owner.
- *domains*: domains are an ensemble of contexts. Every context has got a series of services and making more contexts share the same structures is a noteworthy economic saving. Contexts can securely communicate inside the same domain. Domain policies are established by the "domain authority".

Every principal can define some policies, though there is a hierarchy that must be strictly observed, that is to say:

- Domain Authority;
- Context Master;
- Aglet Owner;
- Aglet Manufacturer.

It means that policies defined by the domain manager can be applied or modified by the context master or other principals. The context authenticates the owner of an aglet that still has to be launched. Later the aglet is launched and the owner defines its security preferences. Then the context instantiates the agent, after having included information concerning the owner, the manufacturer and the source context (in this case itself). Such information, together with the agent's code and the owner's preferences, are signed by the context and become the static

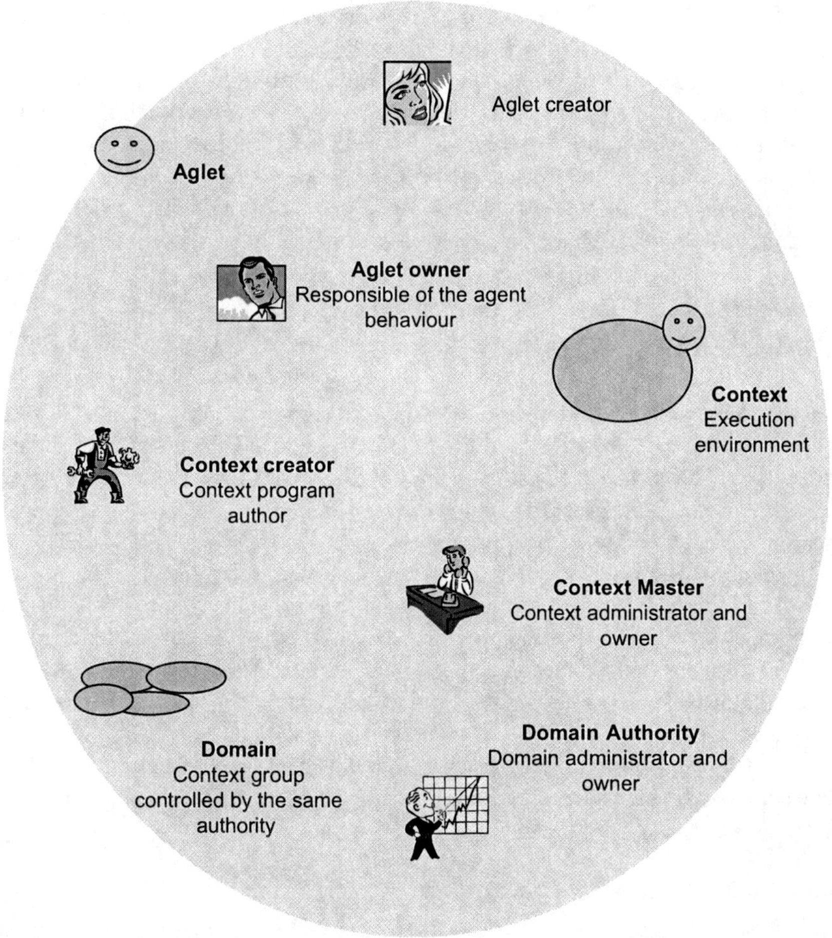

Figure 7: Principals in aglets security model.

part of the agent. Every context receiving the agent can verify the integrity of the agent's static part.

The agent's itinerary is described as follows:

- source context (the context that has created the agent);
- target context;
- current context (that receives the aglet travelling towards the target context).

Between the current context and the target one a safe channel is established; a protocol with client authentication is actually used. The aglet exploits the new context to have information on the new environment, for example to know whether there are other agents to interact with them: access to resources occurs according to the policy established by the context owner. Authorizations are released if the agent owner and/or the manufacturer have been authenticated, according to the kind of aglet, the calculation power, the occupation level, etc. In accordance with the Context Master lines, the context grants permits to obtain information, read, write, delete files, connect to a network port or to another context, load a library, create new aglets, etc. The aglet owner's preferences are taken into consideration only if they do not interfere with the context policy. Finally, the context constantly monitors the agent's use of resources.

10 Monitoring and security

Keeping under control the operations done by agents is necessary not only to check mobile agents' access to resources, but also to check the resources used by agents identified at the execution moment, for example to protect against possible attacks. For the sake of security it is necessary; therefore, to adopt a system that monitors the state of the agents (*monitor*).

By monitor is meant a system, distributed or not, software or hardware or both, gathering information, through suitable sensors, on the state of the target system and, therefore, providing interpretations of them. Probes, or sensors, react to state variations in the monitored system, variations notified to the monitor that, therefore, must give an interpretation, often under the form of Events, and react consequently.

An extremely important particular is the decision on the abstraction level to be used, i.e. defining which class of events will be of interest, in other words the detail level of the application analysis.

10.1 Monitor in an agent-based system: MAPI

As already said, the main reasons why software companies cannot completely trust mobile agent technology yet are due to likely problems stemming from hosting "alien" active entities. Research has gained remarkable results in the fields of both mobile agent security and host sites protected against agents' execution [13].

Nevertheless, trust in mobile agent systems can only be earned by incorporating some components to carry out measurement as well as monitoring, and eventually limit the amount of resources used by the agents themselves during their execution. The availability of a component for on-line monitoring of mobile agents is thus crucial for every agent platform, in order to control their behaviour during their execution and to enable a correct manipulation of the operations managing the agents' use of resources.

The *Monitoring Application Programming Interface* (MAPI) is a monitoring component for Java-based mobile agent environments. Let us remember that many platforms adopt Java as implementation language, since it has clear advantages in mobile agent support (dynamic load of classes, serialization, security and uniformity mechanisms). However, the JVM tends to hide platform-dependent information and is an obstacle to on-line monitoring. That is why some extensions of Java technology helping solving this problem have been introduced: *JVM profiler interface* (JVMPI) [20] exports several JVM inner events for debugging and monitoring, while *Java native interface* (JNI) [21] makes it possible to integrate Java programs with some code executable in dependent platforms. Moreover, Java structure can be integrated with components and tools compatible with the most widespread Internet protocols, such as SNMP [22].

MAPI has been integrated with the *secure open mobile agent* (SOMA) platform [23]. Such component exploits JVMPI to gather various-level events produced by Java applications (objects allocation and methods invocation). In addition, it makes use of JNI technology to make possible the integration with platform-dependent monitoring native mechanisms, which are currently implemented by Windows, Solaris and Linux. MAPI also encapsulates the components SNMP and filters their information to aggregate monitoring indicators; thus, it limits the overhead due to long sequences of SNMP requests/responses.

10.2 Technologies for Java-based mobile agents on-line monitoring

Since many mobile agent platforms are implemented in Java, we can classify resources into "*JVM resources*" and "*OS resources*". The first ones are defined as resources supported by JVM and visible to Java-based execution agents; the second ones are execution resources of the Hosting Operative System, and are on a lower abstraction layer than the first ones. Java-based mobile agent platforms can exploit mechanisms provided by Java programming environment to control the access to JVM resources. They usually inherit the advantages of the *Security Manager* that controls the accesses to JVM resources, by defining access permit classes through suitable security policy. However, it can only check Java code access permits, while mobile agents need also other mechanisms and tools able to dynamically verify use quality in access to JVM and OS resources.

There are two possible solutions. The first one takes into consideration the creation of an *ad hoc* resource manager for every kind/case of resources potentially accessible by mobile agents, and to oblige agents to work on resources only by passing them to intermediate entities. In this case, agents do not have direct visibility and must always use some resource management mediators. This solution needs the writing of at least a manager for every kind of known resource, and also imposes a proxy overhead for every access. The second possibility consists in managing an on-line monitoring service inspecting JVM and OS resources state during the agents' execution. That can be made by using JVMPI and JNI technologies, which makes visibility possible without JVM, and by integrating them with external and standard SNMP monitoring components.

10.3 Local monitoring and mobile agent control in SOMA

As we have already seen, SOMA is a programming structure designed and implemented to support the opening of services in an Internet environment [23]. It has a modular organization in a set of middleware services and is implemented in Java to achieve certain portability (fig. 8).

SOMA provides services for the agents' migration, naming, communication, security and interoperability. In order to carry out scalability, which is a crucial aspect in the Internet global scenario, SOMA offers locality abstractions to describe every kind of interconnected system. Every node hosts at least one region for the agent's execution. Various regions are grouped in branches abstractions matching local communication networks. In every branch, a default region is charged with functional starting branches, integrated with components via CORBA. The mobile region is the locality abstraction used to support mobile devices: it increases the number of local regions having specific features for automatic reconfiguration when attachment domain changes [24].

SOMA exploits MAPI to obtain information monitoring about the resources used by mobile agents, both local and remote ones. If MAPI is responsible for local indicators, SOMA has the visibility of distributed resources' global state through *ad hoc* use of monitoring mobile agents. They are charged of gathering, filtering and carrying required data in order to build a distributed network using monitored information as the basic knowledge to apply suitable administrative policy. Monitoring indicators are used to control and deny specific mobile agents operations on available resources. For instance, in case of local overload, SOMA administrators can dynamically limit the number of operations available to agents authorized by remote users.

MAPI monitoring introduces some overheads on the different supported operating systems. Overheads depend mainly on monitoring indicators and on the intervals between updating; SOMA administrators, in response to the service/system's execution time conditions, can dynamically set them up. It is significant that MAPI does not impose any change to standard JVM. That can be considered

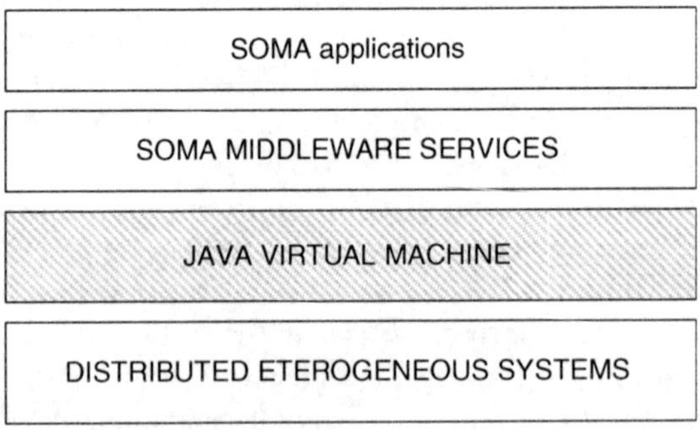

Figure 8: Modular architecture of SOMA platform.

a basic feature of any platform based on mobile agents aiming at working in global and *Open-Distributed* environments, such as the Internet.

Fig. 9 shows MAPI architecture that provides a uniform monitoring interface, independent from the platform's heterogeneities. It is implemented by the *Resource Manager* class integrating three different components: MAPI *Profiler Agent*, MAPI SNMP *Agent* and *MAPI*ResManager*.

MAPI *Profiler Agent* is able to receive monitored information about JVMs and the state of resources. It gathers JVMPI events, filters and processes them on-line, in order to offer concise monitoring indicators during the service execution. Monitoring functions based on JVMPI are immediately transportable on every host executing JVM's version 2. According to SNMP terminology, *MAPI SNMP Agent* acts as an SNMP administrator questioning SNMP standard agents available on a target, in order to obtain the monitoring of OS local resources data. The *Resource Manager* can exploit *MAPI*ResManager* classes in order to carry out OS resources visibility by integrating it with platform-dependent monitoring functions, through JNI.

A basic attitude in carrying out MAPI tool is not to change JVM standard. That choice, together with the possibility of examining mobile agents without intervening on their source or executable code, is fundamental to adopt MAPI as a distributed on-line monitoring tool, by examining the services of mobile agents based on the Internet open infrastructure.

10.4 Distributed monitoring

MAPI's components also constitute the basic mechanism that provides SOMA's distributed administration: it examines and filters recovered data, and coordinates implemented monitoring local entities, according to their itinerary; it also examines and controls distributed systems and services through SOMA mobile agents.

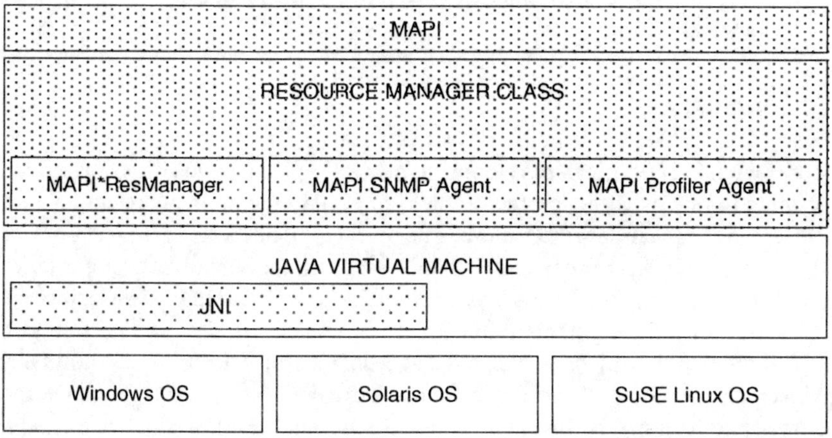

Figure 9: Java-based MAPI architecture.

In order to carry out SOMA mobile agents monitoring and distributed control, two kinds of intercommunicating mobile agents have been created and implemented: *managers* and *explorers*. Every *explorer* mobile agent is charged of gathering data by monitoring a series of targets (i.e. target domain) usually belonging to the same local network. The *manager* mobile agent, on the other hand, commands the explorer agent, links up the results found by the explorer mobile agents, examines them and shows a global perspective of visited domains and of achieved goals (executed operations) to the system's administrator. Moreover, the manager can delegate some of his operations to the explorer agent.

Various kinds of organizations are possible, with different hierarchy levels and mobile agents number, both manager and explorer, adopted for target domain.

11 Future scenarios

Even if execution mobile agent operations monitoring and checking find wide acceptance in agents technology, there are still unsolved problems concerning mobile agents, above all as far as the creation of suitable platforms is concerned. There is no mobile agent programming environment yet, offering mechanisms for measuring, controlling and describing the agents' resources without imposing changes to JVM standard.

Many research activities are in progress, to better explore the possibilities of exploiting mobile agents as suitable technological implementations for the monitoring and management of distributed systems and services with a high degree of security.

The first results of the monitoring and checking of resources used by SOMA mobile agents have shown the flexibility of Java-based on-line monitoring. However, encouraging such results may be, a lot of work has to be done yet in order to carry out a complete and flexible control of the resources used by SOMA mobile agents.

12 Conclusions

Mobile agents offer great development possibilities to applications based on distributed systems, although the problem of how to maintain a certain level of security hinders their large-scale use. The protocols dealt with in this section are only a part of the research conducted in this field; still they are of great importance since they deal with problems not easy to be solved. Actually, ensuring the agents' integrity from potential attacks is essential for them to move freely in a distributed environment. Meanwhile, the defence of execution environments against malicious agents is fundamental to ensure the development of technology: the owners of computers forming the network will never allow an alien element, such as a mobile agent, to be executed on their machines until they receive enough guarantees that the agent is not able to damage the computer.

Let us notice, finally, that in almost the majority of adopted techniques, cryptography and digital signature play an important role in the exchange of safe information, also thanks to the presence of "third trusted parties" whose presence is fundamental if heterogeneous entities such as mobile agents are used.

References

[1] Adleman, L., Rivest, R.L. & Shamir, A., A method for obtaining digital signatures and public key cryptosystems. *Communication of the ACM*, **21(2)**, p. 15, 1978.

[2] Fugini, M., Maio, F. & Plebani, P., *Sicurezza dei sistemi informatici*, Edizioni Apogeo: Milano, 2001.

[3] *Advanced Encryption Standard,* FIPS PUB 197, National Institute of Standards and Technology, Secretary of Commerce, November 2001.

[4] Zunino, R., *Introduzione alla crittologia: protocolli*, DIBE Università degli studi di Genova, 2000.

[5] White, G.B., Fish, E.A. & Pooch, U.W., *Computer System and Network Security*, CRC Press: Boca Raton, 1996.

[6] Tanenbaum, A.S., *Computer Networks*, 3rd Edition, Prentice-Hall Inc.: Upper Saddle River, NJ, 1996.

[7] Diffie, W. & Hellman, M.E., New Directions in Cryptography, *IEEE Transaction on Information Theory*, 1976.

[8] Schenier, B., *Applied Cryptography*, 2nd edition, C. John Wiley & Sons: New York, 1996.

[9] Giustozzi, C., Monti, A. & Zimuel, E., *Segreti Spie codici cifrati*, Edizioni Apogeo: Milano, 1999.

[10] Jansen, W.A., *Countermeasures for mobile agent security, Computer Communications*, special issue on Advanved Security Techniques for Network Protection, Elsevier Science: Amsterdam, 2000.

[11] Hohl, F., A framework IO protocol mobile agents by using reference states, *Proc. of the 20th International Conference on Distributed Computing Systems* (*ICDCS*), 2000.

[12] Hohl, F., A new protocol protecting mobile agents from some modification attacks. *Technical Report Nr. 09199, Faculty 01 Informatics,* University of Stuttgart, Germany, 1999.

[13] Corradi, A., Cremonini, M., Montanari, R. & Stefanelli, C., Mobile agents integrity for electronic commerce applications. *Information Systems,* **24(6)**, 1999.

[14] Park, J.Y., Lee, D.-I. & Lee, H.-H., Data protection in mobile agents; one-time key based approach, IEEE, 2001.

[15] Riordan, J. & Schneier, B., Environmental key generation towards clueless agents. *Mobile Agents and Security,* ed. G. Vigna, *Lecture Notes in Computer Science 1419*, Berlin, 1998.

[16] Mell, P. & McLarnon, M., Mobile agent attack distributed hierarchical intrusion detection system. *Proc. of the 2nd International Workshop on Recent Advances in Intrusion Detection*, West Layfayette, IN, 1999.

[17] Neuenhofen, K. & Thompson, M., A secure marketplace for mobile java agents, *Proc. of the 2nd International Conference on Autonomous Agents*, Minneapolis, Minnesota, USA, ACM, pp. 212–218, 1998.

[18] Hercock, R.G. & Gifford, I., Solutions to security in mobile agent systems, available at http://ieeexplore.ieee.org/lpdocs/epic03/, 2001.

[19] Karjoth, G., Lange, D.B. & Oshima, M., A security model for aglets, *IEEE Internet Computing*, **1(4)**, pp. 68–77, 1997.

[20] Java Virtual Machine Profiler Interface (JVMPI), available at http://java.sun.com/

[21] Gordon, R., *Essential Java Native Interface*, Prentice-Hall Inc.: Upper Saddle River, NJ, 1998.

[22] Stallings, W., *SNMP, SNMPv2, SNMPv3, and RMON 1 and 2*, Addison Wesley: Boston, MA, USA, 1998.

[23] Bellavista, P., Corradi, A. & Stefanelli, C., Protection and interoperability for mobile agents: A secure and open programming environment, *IEICE Transactions on Communications*, IEICE/IEEE Special Issue on "Autonomous Decentralized Systems", **E83-B(5)**, pp. 961–972, 2000.

[24] Bellavista, P., Corradi, A. & Stefanelli, C., Mobile agent middleware for mobile computing. *IEEE Computer*, **34(3)**, pp. 73–81, 2001.

Data mining and information retrieval

A. Benanti, G. Capodici, S. Greco Polito, F. Nuccio, R. Scarvati, S. Sorce and A. Genco
DINFO – Dipartimento di Ingegneria Informatica
Università degli Studi di Palermo

1 Introduction

Data analysis is a classical problem which constantly keeps on offering new challenges: the birth of distributed computing models, based on networks of calculators has been establishing a new, extremely important dimension to this problem.

At the present time, the statistical modelling, the automatic learning and the techniques of knowledge detection are mainly employed in a centralized way; this approach asks for all the information from several sources to be stored in the same physical place. This fact often brings to remarkable costs in communication; it takes a long time of answer and opens a background of security problems. Particularly when the information sources are heterogeneous (such as sites having data with different characteristics), the classical statistics and the automatic learning must perform a preliminary and fundamental operation: moving the data to the central site, where the elaboration will be made, before any analysis.

Even though during the last years the available communication band has become broader and broader; this growth does not have the same speed of the increasing information availability. Therefore, downloading great amounts of data towards only one site, on one baseband channel, followed by the application of centralized algorithms of data analysis, cannot be a scalable solution for future applications of data analysis coming from big distributed information sources. In most cases, it also could not be feasible for problems of security, privacy or simply for incompatibility at a level of representation of the information.

Thus, it has been necessary to reformulate the fundamental approach to the analysis of the distributed data. As known, the mobile agents have many advantages in searching for applications of the distributed information. By drawing from a third information agent, a mobile agent can call upon the local operating resources, removing the transfer on the net of intermediate data.

On the other hand, the fact that an agent can keep on being executed even if the net is made up by indirect and unreliable connections causes the mobile agents to be particularly suitable to work in mobile calculation environments.

The most important thing is the possibility to carry out complex, efficient and steady behaviours, with a very short code. An agent can, in fact, choose different migration strategies depending on the task entrusted to him and on the current conditions of the net. Therefore, it is able to change its strategies of search and of the data analysis as soon as the net's conditions change.

Even though each of these advantages is distinctive of the mobile agents, no one of them is exclusive of such paradigm.

Each specific application, indeed, can simply, effectively and in a steady way be implemented by more traditional techniques than the mobile agents. Since different applications require several traditional techniques, in many applications, a combination of them is often necessary.

In brief, the real strength of the mobile agents is not the possibility to build up new distributed applications, rather such paradigm allows an easy implementation – strong and efficient at the same time – of a wide field of distributed applications inside a single system.

The mobile agents have several points of strength which make them particularly attractive in applications of distributed calculation where there are both a low bandwidth and a low net reliability:

- in order to migrate towards the desired location where the resource is contained, an agent can interact with the resource without transmitting intermediate data through the net, reducing, in such way, both the band occupation and the latency;
- when the interaction between a mobile agent and user takes place in "local" it is quicker;
- the user can anyhow continue his interaction with the resource through the mobile agent, even though the direct connection is temporarily interrupted;
- the mobile agents allow the traditional client and servers to move the work activities indifferently from one to the other depending on the computing capacities and on the current workload;
- the most widespread distributed applications can be naturally fitted in the mobile agents model since a mobile agent can sequentially migrate through a set of calculators, or it can generate child-agents in order to visit more computers at the same time.

Even though each of these advantages characterizes the mobile agents, no one of them is exclusive of such paradigm.

Each specific application can be effectively implemented through other techniques: e.g. message-passing, calls to remote procedures, remote invocation of methods-of-objects, RPC (remote procedures call) queuing (the RPC calls are queued for a next invocation if the net connection fails), Java applets and servlets (that are Java programs which are downloaded by a web browser or uploaded to a web server), automatic installation services, query languages for specific applications, etc. None of such techniques however contains all the mobile agents' advantages.

In order to make the mobile agents' paradigm interesting in a wide range of applications and particularly in the data mining field, it is necessary to face two key-problems:

1. Mobile agents must become more scalable: the principal problem of the scalability lies on the performance of the agent's low-level infrastructure. In particular, the overloading intercommunication among agents has to be reduced. By adopting this solution, the stationary agents can compete with traditional client–server implementations. The overloading of net due to agents' migration has to be reduced too. In such way the migration will be favourable for the agent also in the best net environment; even it only needs to invoke few operations for each information resource.

   Finally, the execution environments must be able to execute the agents as if they were compiled in native code; in such way the agents can be, then, used for tasks of workload balance. Solutions to such implementation problems exist both in the high-performance servers and in the mobile agents literature, and the principal task is to identify and apply the most suitable solution.
2. Mobile agents require a lot of information in order to take some decisions on when and where to migrate. Several support services are necessary to obtain and analyse the current state of the net, of the target calculator and of the information repository conditions: all that in order to have an efficacious action-planning to be executed, to reach the aim wished. Some of these services have been developed in the sphere of the distributed calculation, but it takes a lot of work to make such systems interact with the mobile agents, in which the software components move quickly and continuously from one PC to the other. Other services are more exclusive than the systems of mobile agents.

   Such services include planning-algorithms which allow a single agent, or a little group of cooperating agents, to identify the best migration path through the net; other algorithms allow, for instance, a mobile agent to determine a good observation strategy of a collection of documents changing in time, etc.

The World Wide Web has established itself as an interactive mean for the collection of the information always becoming more and more popular. The vastness of the Web, the heterogeneity of the calculators connected to it, the various architectures of data storage and the speed through which these can change have induced the research of data mining architectures, able to work in heterogeneous environments and to return useful information in real time. It follows the birth and the development of active forms of data mining which have been added to the passive classical ones.

Figs. 1 and 2 show the differences between the two kinds of data mining mentioned above.

From the figures you can observe that to each node of the net it is possible to associate a data mining unity constituted from a data object that identifies a partition of the whole database, and by a program object.

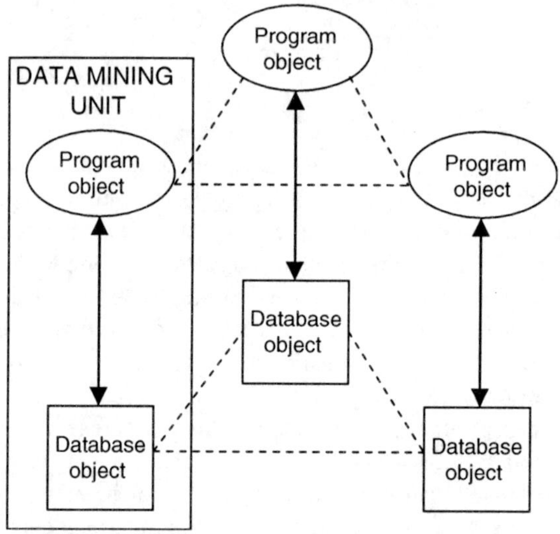

Figure 1: Passive data mining.

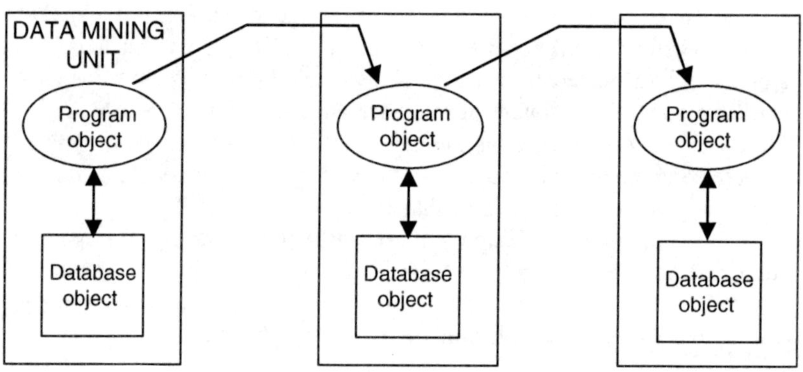

Figure 2: Active data mining.

In the passive form, for each node, a copy of a program object exists, which can only work on the database resident in the same unit, while in the active form only one program object exists, which moves from a node to the other of the net, working on current data object resident on the data mining unit.

For the implementation of active data mining we could invoke mobile agents paradigm, using their capacity to preserve their own state during migration from a node to the other on the net, and to add to their code the useful information extracted during local execution of the databases in the nodes previously visited. It is possible to observe so far how the capacity of the agent to act locally avoids the not-useful information transport from the server to the client and considerably reduces the communications on the network compared to a traditional mining system, such as the RPC.

2 Design and implementation of a data mining system

Incorporating a new algorithm of automatic learning inside a system does not imply only taking into consideration all the problems born by the compilation of external software, but also the realization of a mapping that goes from a generic functionality of automatic learning to the implementation of a set of calls to functions which implement the particular algorithm.

All the algorithms of automatic learning follow a basic principle: in Ref. [1] it is affirmed that it is possible to define few generic calls to the tasks provided by the service of automatic learning.

Such generic calls will constitute the basic interface, common to all the algorithms of automatic learning system.

Each knowledge discovery and data-mining (KDD) configuration process must be real time executed, that is, the system must be able to execute the data mining task entrusted to him choosing, in autonomous way, the most adequate automatic learning algorithm.

It is possible to guess how to distinguish two levels of a learning: the specific level of learning is the one realized by the appropriate data mining task; the general level is a meta-learning used to finish the process through which the most adequate data mining algorithm is decided for each session of elaboration of the information. The meta-learning process is not banal, as there are many details to be considered, some of which are:

- *Available resources*: the choice of a particular algorithm should be bound to its own adequacy to be assigned to a particular machine at a given time (depending, for example, on the amount of current work of the CPU, the available memory, etc.).
- *Available band*: in a non-parallel and distributed KDD system there is a relevant bottleneck given by the big amount of data to be treated. The approach could be that of moving data towards the machine in which the chosen algorithm is in execution, so neither the task is parallelized nor the code is moved.
- *Accuracy of the discovered information*: the algorithms of automatic learning differ from one another for tolerance to the noise and efficiency in the production of new knowledge. There can be users who want more accuracy than efficiency or vice versa.
- *Ways of structuring the discovered information*: there are many algorithms of automatic learning, and the topology of a newly produced information depends on the algorithm actually used. Some algorithms produce trees of decision, some a files classification and others quantitative relations among variables. New knowledge format can be imposed by the user.

It is necessary, therefore, to associate a set of characteristics, which shows the requests for resources and the quality of the service to each specific implementation of an automatic learning algorithm.

3 Data collection with mobile agents

Let us consider the following situation: there is a central server which must pick up data from many computers: a web search engine which collects data from the web server of all the world; a central information server of a company which picks up data from various departments; a data mining software on a big and distributed database.

There are three different solutions to such problem:

- *conventional data collation*: a central server picks up data from many computers. Such solution has the disadvantage to receive data from all the computers before the elaboration starts.
- *data collation with distributed search engines*: all computers execute a type of distributed search engine, such as Harvest. Local search engines elaborate data at a local level and move the results to the central server. Such solution has the advantage to have low net traffic, since the local computers execute the elaboration before moving data to the central server; but they also have few disadvantages: (a) a lot of maintenance of the local search engines; (b) when a new version of the search engine is released, it must be installed in each local server; (c) it is not a very flexible solution when the research algorithm has to be changed.
- *data collation with mobile agents*: a mobile agent goes through the net and processes data on each computer, sending the results to the central server. Such solution has the advantage to produce low traffic, since the agents locally execute the processes; when the research algorithm changes a new agent is sent and therefore no manual updating is necessary on the remote servers; it is furthermore usable in not very reliable environments. A problem generated by such solution is that the agent's executable code must be moved, thus yielding to the occupation of a significant net band part, unless there is a shared code in the host's agent. To avoid this problem, some techniques such as shared packages or caching can be useful. It is furthermore useful to move also the state of the agent.

4 Request for information and proxy caches

The proxy caches are tools which allow the acceleration of the process of information extraction; most frequently requested web pages are preserved in them. Their use allows the decrease of the time of answering and avoids the congestion of the servers which, in the last few years, with the increase of requests for access, has become a not-negligible problem.

The principal problem for proxy cache management is the consistency of data contained in it; it is necessary that these "copy data" follow the changes of origin data.

The classical solutions to such problem consists in using some algorithms based on periodic requests to the web servers of the verification of a "change event", but normally these requests travel along the network through long paths with a consequent increase of workload and a decrease of caching accuracy.

Such limitations can be overcome adopting a mobile agents system that allows to realize the poling on minor distances: a mobile agent (MA) can be sent in a node close to the web server (WS) containing the original objects; such agent has the task to follow the objects change and possibly report it to the proxy for the renewal of the copies. The mobile agent can realize the high-frequency poling on the web server without causing an increase in the load on the network because it is placed in a node close to the server itself.

To define a project based on the previous criterions we have to consider a package constituted by a static agent (SA) which evaluates the requests for exit to the local proxy cache (LPC) using the ICP (Internet caching protocol), and possibly sending mobile agent to the nodes containing the original objects.

When the mobile agent informs the static one of a "modification event" occurring, it restores the LPC through an HTTP request.

We intentionally presented the not-optimum solution, but wide application ray one, that is a solution which can be applied also on servers which do not support mobile agents through HTTP requests. Optimum solution would have been obtained directly sending the mobile agent on the node containing the original objects.

5 Route planning

A more general class of IR problems anticipates the possibility of an agent to be unable to find the wished information on a destination computer.

According to what has been previously affirmed, it is hoped that the net resources will be used as less as possible. Such target can be reached sending a lower number of agents than the amount of possible destination computer. In this case we have to plan the best sequence of machines each agent must visit, that is establishing the route, in such a way that the wished information can be found as soon as possible.

A route determined by the planning is correlated to:

- a list of CPUs in which an agent can find the wished information;
- the uncertainty on quality of the available data on such CPUs;
- the current conditions of the net.

The list of CPUs and the uncertainty on informative quality of the documents can be supplied by an advanced version of yellow pages service.

The uncertainty degree is defined by the probability with which an agent can successfully find the information in each of such CPU.

At last the net conditions include the information which concerns link connectivity, CPU operability on the net, latency and link's band availability. These statistics are picked up by an appropriate form for the net monitoring.

As well known, to determine whether the system which uses mobile agents bears significant benefits to distributed applications compared to the traditional approach is a still open matter. It is necessary to consider a lot of parameters to evaluate the performances of the paradigm used in the development of distributed applications. The performance is influenced by both the chosen paradigm and the strategy of migration used by the paradigm itself. Therefore, a quantitative model is necessary to decide an appropriate interaction in the development of distributed applications.

In Ref. [2], Chan *et al.* suggested a model of evaluation of the performances for mobile agents compared to the RPC paradigm that is the traditional architecture client–server.

The planning system shown in Ref. [3] consists of three principal components: a planning form, a form for the net monitoring and a yellow pages form. When a mobile agent is asked to find an information, it first consults the planning form which asks, therefore, the yellow pages one for the possible locations in which the mobile agent can find its wished information.

It is supposed that the yellow pages form is able to measure the success probability that we assume to be quantifiable (e.g. ratio among data stored in a proxy server and the total amount of available data in the current server).

After obtaining the list of the CPUs and of their correspondents success probability, the planning form gives the list to net analysis form, which gets the latencies, the bands between the CPUs and their current CPU load back. The net analysis form keeps track of such statistics analysing the net according to pre-established intervals.

As soon as the net statistics get back to the planning form, the sequence in which the agents must visit the nodes (or minimize the total time of execution expected) is calculated on the net statistics and on the success probabilities, using algorithms similar to those that solve well-known travelling salesman problem.

Because of the complexity of such a problem, a few simplifications to easily obtain excellent solutions can be assumed in Ref. [3].

They have concerns with:

- number of the mobile agents
- net latencies
- success probability
- calculation time slot for each CPU

The complexity of the problem of the "travelling agent" can be reduced assuming that the latencies among nodes are all the same. For example, if the time slot of calculation in each node is extremely high compared to the latency among nodes, the differences between the latencies can be ignored or even be assumed

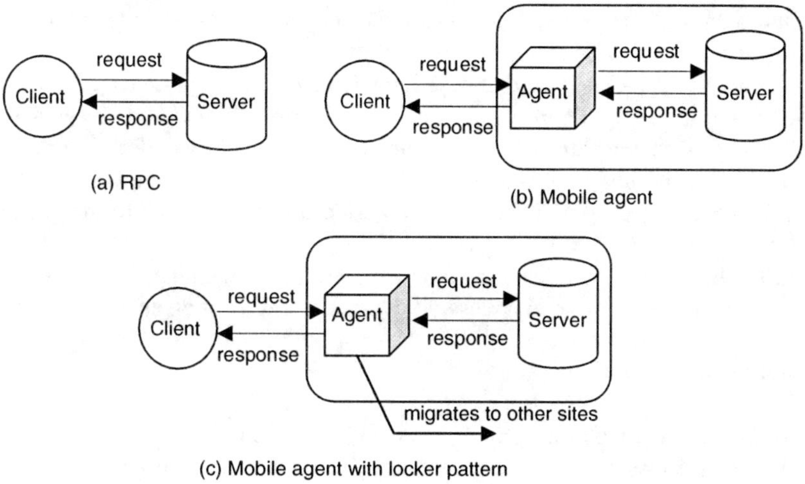

Figure 3: Three paradigms of distributed computing.

equal to zero. Alternately, if there is no information about inter-node latencies, it is possible to assume them all constant.

Chan *et al.* in Ref. [2] considered three types of pattern interaction: RPC (client–server traditional architecture), mobile agents and mobile agents pattern Locker (fig. 3).

In Locker, the mobile agent temporarily stores the data privately. In such a manner it can avoid bringing the data which at the moment are not necessary. Afterwards, the stored data can be sent to the customer node, or another mobile agent migrates to the nodes themselves in order to pick up the data stored in private areas.

The performance of each paradigm is based on the value of a few parameters. The principal ones considered by Chan *et al.* [3] are band of net, dimension of picked-up data, number of interactions with the server, mobile agent's dimension and so on.

Using the mobile agents' paradigm a short time slot is necessary to realize a net connection, while a longer time slot is necessary to move the mobile agent together with picked-up data at its inside in the information retrieval phase.

According to the RPC paradigm executing a high number of remote communications will produce great delays; meanwhile, a mobile agent needs less time to migrate on the net, since the code to be moved is usually lower than the amount of the data to be treated.

So, a mobile agent shows a better performance than the RPC approach when it requires frequent global communications or when the condition of a given net is good. On the contrary, the RPC shows better performances than the mobile agents approach when it requires less remote connections or when the particular condition of one of these is not of good quality.

The number of global communications depends on the number of resident data present in each node in case of data mining applications. Mobile agent and Locker approach's performance is influenced by mobile agent dimension. The evaluation model just presented shows that there is no paradigm that is better than the others in absolute: the choice must be made according to the type of application.

Particularly it is noticed that execution times on the net change proportionally to the sequence of the visited nodes and interaction mode of given paradigm.

5.1 Observing agents

The above-described planning methods are all correlated to the information given by the advanced yellow pages service. Such service provides a probability of success in the research field of certain types of information resident in the given nodes. The yellow pages must be kept up to date to add new sites, to remove the old ones and to re-index those which have been changed. Some yellow pages can lead the researcher to general sites of documents, while some others can lead to more specific ones, such as web pages.

To solve the indexation problem, the limited calculation, the net and the storage resources must be able to assign a score to the new documents collation, or re-check the old ones to inspect them, in order to look for changes.

Whether the task is done sequentially or in parallel, a search engine must be able to decide which document has to be examined next.

Once the knowledge of the history of the document is given and the priority coming from its quality level is assigned, there are several problems to be dealt with, such as the best moment to re-check the document and the best way to describe its state.

If the resources were limited, the answer would be simple: each of the documents should be monitored, looking for changes, as frequently as necessary. A dynamic observation of the data stored in a site has obvious associated costs such as timing of the node activity to recover and examine the data (net latency and cycles of CPU) and necessary memory to store the results, etc. For such cost, the search engine takes advantage of having a more up-to-date index of previously explored documents, a more complete collection (if new documents are known) and a more careful representation of the dynamics of the document in question, that is, how it changes from time to time. A comprehension of the modes with which the documents change in time, in order to maximize the updating of such index, is necessary. The knowledge of this dynamics of changing data allows to do observations more in conformity with the reality.

When the search engine must decide the document to examine, only some of them are likely to be changed since the last analysis. If the aim is to look for content changes, it has more sense to re-examine only the documents that quickly change rather than considering those having greater stability.

6 Performance evaluation

Krishnaswamy *et al.* [4] suggested a hybrid model for the distributed data mining (DDM), planned to meet the DDM systems requests, to work in e-commerce and application service providers (ASP) environments (fig. 4). The main characteristics of such architecture are the integration of the client–server model with mobile agents paradigm and an optimizer which provides the estimate of the costs for DDM tasks.

The system is able to complete data mining tasks on remote sites by the use of mobile agents, including also some data mining servers endowed with well-defined calculation resources. This characteristic helps the system to satisfy the various necessities of clients that can be extremely heterogeneous among them. The costs' estimate refers to problems of cost and optimization of the DDM processes.

Krishnaswamy *et al.* [4] show a DDM in an electronic commerce environment and its hybrid system:

In a hybrid system, the users request for data analysis services by the connection to data mining server, which is a high-calculation-performance one that works both as controller of the distributed data analysis process, and as a calculation-dedicated-resources provider. The server has in its inside the distributed data mining management system (DDMMS) which executes the several tasks associated to the DDM process.

The DDMMS is the fulcrum of system architecture and includes (fig. 5): the "*user manager*" sees to the users' authentication and to identify the data mining task to be executed; the "*algorithm manager*" is responsible for the maintenance of the algorithms which are the components of data mining system: each user can, indeed, register his algorithm inside the system, and can decide to make

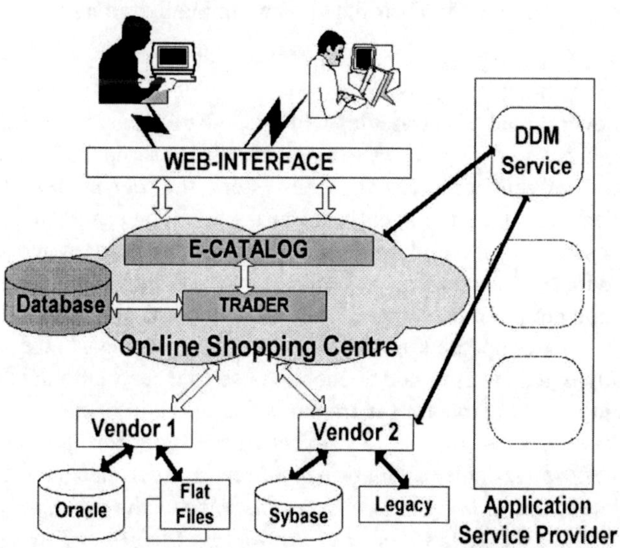

Figure 4: Distributed data mining in an e-commerce environment.

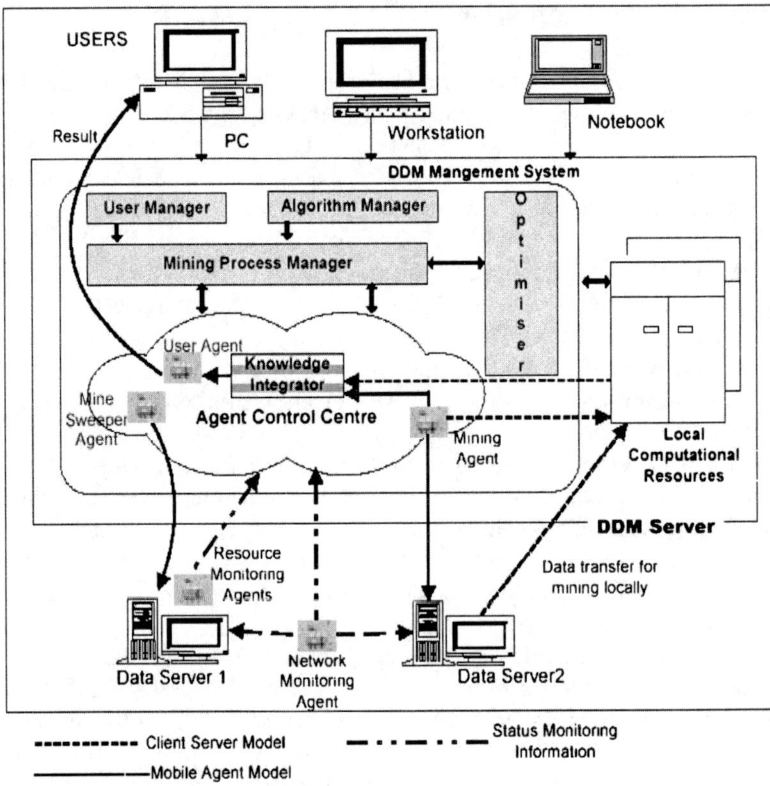

Figure 5: Distributed data mining hybrid model.

it visible to others; the "*optimizer*" is responsible for calculating the estimated cost for several analysis strategies which can be used by turns to meet the user requests; the "*mining process manager*" supplies the coordination services among the different components of the system; the "*agent control centre*" is the platform inside which the agent's activities take place; the "*user agents*" provide the users with the updating on the state of the task entrusted to them and with the data analysis final results; the "*network monitoring agent*" analyses the connections among data servers, continuously exploring the net and therefore keeping the states updated in the "*mining process manager*"; the "*data resource monitoring agent*" is assigned to each data source becoming part of the system: such agent is responsible for the transmission of the information concerning the data source contents; the "*mine sweeper agent*" goes through the data servers, determines the available calculation resources and estimates the amount of data to be treated; the "*mining agent*" is an instance of the data analysis algorithm decided to be executed; at last the "*knowledge integrator*" merges the results obtained by the different information' sources and gives the final result to the "*user agent*" which communicates it to the user.

6.1 Mobile agents model

The situation to be faced when a data mining problem – divided into several information data sources – is examined, is the following: in a distributed environment where it is necessary to execute a given data mining task, it is suggested to use the mobile agents' paradigm.

The principal steps to be followed are: the request for execution of a task by an user; the delivering of the mobile agents to the respective data servers; the sheer data mining phase which takes place "in local" and the coming back of mobile agents from the data sources together with the final results of the led analysis.

The model described above is characterized by a set of mobile agents that pass through the various data servers, that is the different information sources, to complete the requested analysis.

In general, such concept can be expressed considering m agents passing through n sources. According to such simplification there are three possible alternatives:

- $m = n$: one data mining agent for each involved source;
- $m < n$: it will be asked to a few agents to pass through more than one server;
- $m > n$: it is equivalent to the first case.

Each of the previous alternatives presents a cost function.

6.1.1 Case $m = n$ (or $m < n$)

The algorithm used for the server of different kinds of data can be diversified or uniformed. The system sends a mobile agent encapsulating the data mining algorithm (with a few parameters) towards each of the data server that takes part in the DDM activity. The situation is illustrated in fig. 6.

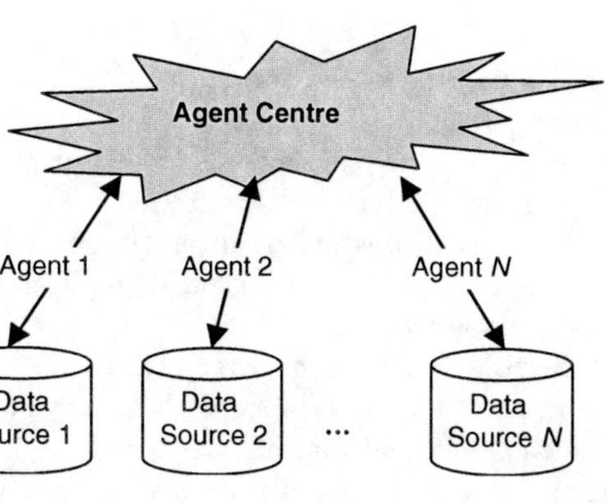

Figure 6: Same number of mobile agents and data sources.

If the i-th source ($1 \leq i \leq n$) has to be analysed, the cost function for the answering time is

$$tddm = tdm(i) + tdmAgent(AC, i) + tAgent(i, AC)$$

in which

- *tddm* is the answering time necessary to accomplish the DDM (that is a data server and a mobile agent);
- *tdm*(i) is the execution time of the data mining algorithm;
- *tAgent*(x, y) is the time slot used by the agent to go from the x node to y node. In general the required time depends on the agent's dimension and on the available band among the nodes.

The agent's dimension is expressed in Ref. [5] as: <*State, Code, Data*>, in which "*State*" is the agent's execution state, "*Code*" is the program encapsulated inside the agent that executes the functionality of such program and "*Data*" are the data transported by the agent (as result of a calculation or additional parameters required by the code of the agent).

In the data mining context it is possible to express the agent's dimension as

<*dmAgent state, dmAlgorithm, input parameters*>

The time required by the agent to get back with the results cannot be determined beforehand, since the quantity of the results depends on the characteristics of data. When there are n agents and n data sources, the first ones encapsulate data mining algorithms and parameters (one for each source) and they will be sent on the net by the agents' centre. Afterward the analysis for each source is executed in parallel and the results are again sent to agents' centre. The total time is defined as the time required by the slowest server in order to execute the analysis and to get back the results:

$$Tdm = MAX\,(tdm(i) + tdmAgent(AC, i) + tresAgent(i, AC)).$$

6.1.2 Case $m < n$

There is a difference between this case and the previous one. While in $n = m$ the data were analysed in parallel on all sites, some mobile agents are now assigned to more than one site for the analysis. Therefore, each agent must carry the data mining algorithm and the obtained results from the first datasite on his way towards all the servers to be analysed. The agent will go back to the agents' centre only after having executed the task entrusted to him in all the sources it had to analyse. Fig. 7 shows this situation.

The process of travelling implies a growing of the agent's dimension (caused by the increasing amount of the results obtained by each analysed site).

In the case in which there is only one mobile agent for n data source, the agent is sent from the agents' centre to the first of the n sites. From that moment on, the agent completes the analysing process and carries the obtained results to the next

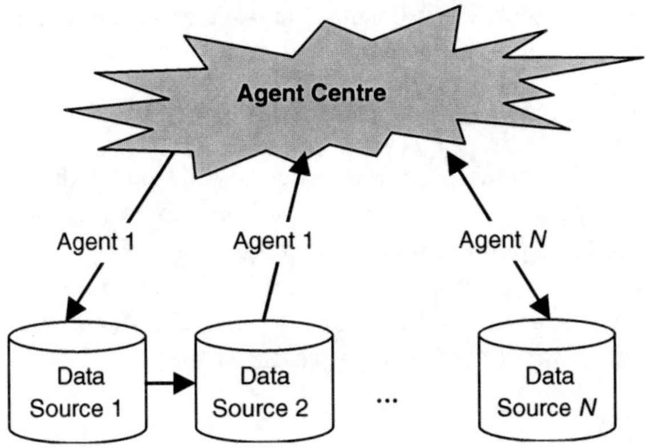

Figure 7: Number of mobile agents minor than the data sources ones.

sites, till each of them will be examined. Thus, the agent goes back to the centre where the integration of the data is executed and then the complete results are delivered to the final user.

The answering time of the DDM is

$$tddm = tdmAgent(AC, 1) + sum(tdm(i) + tdmAgent(i, i+1)) \\ + tdm(n) + tdmAgent(n, AC)$$

and the agent's dimension is specified as

<dmAgentState, dmAlgorithm(s), input parameters + results>

In this case the carried data concern both the input parameters and the results. This implies a growing of the agent's dimension each time it analyses some new sites.

In the general case, to extend the estimating of the costs, *n* data servers can be divided into *m* subset and to the *i*-th mobile agent is assigned the task to analyse the datasites included in the *i*-th subset.

DDM can be executed in a concurrent way (that is *m* different agents work at the same time). To each of these *m* agents that must travel through different sites, a certain number of them can also be assigned in order to be analysed. The total time requested to execute the analysis is the one needed by the agent requiring the maximum time slot to complete the task entrusted to him.

6.2 Client–server model

The estimated cost for the answering time of a DDM system which uses the client–server traditional approach is shown below: the data stored in distributed sources are sent to the data mining server, a parallel and fast one, and then analysed given *n* sources, "*si*" being the *i*-th data subset extracted from *i*-th source. The answering time is given by

$$tddm(i) = tdataTransfer\,(i, DMS, si) + tdm(DMS) \quad 1 \leq i \leq n$$

In this case the data transfer time is added to the time of the analysis: it can be important when the data amount is high and/or there is a narrow band. If data mining is executed in a parallel server, the answering time of the whole process (for all *n* data sources) is equal to the time needed by the data set which requires the maximum time slot; otherwise, if the analysis is sequentially led, the total time consists of the sum of the various necessary times.

6.3 Hybrid model

The hybrid model owns the two main characteristics shown before: it combines the best aspects of the agent model with the client–server approach merging an agent-based framework and a dedicated data mining server.

Such model has the advantage to combine the concept of dedicated data mining resources (and therefore relieving the problems associated to the control loss on the calculation resources, typical of the agents model) with the possibility to go beyond the limits caused by the communication overload associated to the client–server approach. That gives the model the possibility to use the one or the other approach in each DDM task.

Being *n* the number of sites to be analysed, the optimizer operates according to the estimated cost, deciding that *Na* sites will be examined with the agents model and *Ncs* will be examined with the client–server paradigm, $n = Na + Ncs$.

The mobile agents model is obtained when $Ncs = 0$, and vice versa the client–server approach is obtained when $Na = 0$. Assuming that mobile agents tasks and the client–server ones are executed in parallel, the hybrid model answering time is the one requested by the slowest technique.

7 Distributed knowledge nets

Data sources are often geographically distributed. This requires software assistants' use or, more in general, the use of software agents helpful for the collection and for the next analysis of data having, on a large scale, a higher informative content. What is required is a data collation process, both selective and context sensitive.

The distributed knowledge nets (DKNs) are multi-agents organizations, made up of both mobile and stationary agents, planned to support the utilization of heterogeneous data and of distributed knowledge sources, in order to acquire automatic knowledge coming from data and to obtain a support to the decisions.

This leads to the necessity of planning architectures and of algorithms for clever agents able to manage data and sources of heterogeneous and distributed knowledge.

DKNs then, for their nature, must include some tools for:

- the monitoring of different data sources;
- the selective routing of the appropriate information towards specific sites or specific users.

Since the interested information depends both on the context and on the user to which it is addressed, the tools must be adaptable on the basis of the user personal profile and the informative contexts in which we are working.

Since the high-amount information is involved, it is convenient to analyse the data in the sites they are placed in and to transmit only the obtained results. Such consideration explains the reason why the DKNs use software mobile agents able to go through appropriate sites, to elaborate data locally and to feed back the obtained results.

Considering that the data sources are local, that they are elaborated in autonomous way, and that they reside in heterogeneous hardware and software platforms, their real use requires a sufficient interoperability degree among the different information sources (overcoming the intrinsic limits due to the heterogeneity of the involved software/hardware structures): for instance, applications of the public administration must access to data coming from many offices heterogeneous among them. The DKNs use software agents in order to supply an access to such data sources without a solution of continuity.

The data repositories contain heterogeneous types of data (text, images, relational databases, sequences of data, etc.). The DKNs supply tools to extract relevant information from heterogeneous data sources, to change and to assimilate them in a data warehouse in which it can be further analysed in order to make the knowledge discovery process easier.

Data sources are dynamic; they change quite quickly from time to time as soon as a few items are added, modified or deleted, so the DKNs include software agents able to identify and propagate the changes, introducing the necessary modifications in the basic knowledge.

Let us consider, for instance, particular applications in the military field: the interpretation of data coming from satellite images can be influenced also by the contents of reports collated by different sources of independent information.

The high amount of data, the vastness of complex and potentially relevant interrelations that must be known and the diversity of data sources strongly test the data mining approaches and the automatic discovery of knowledge. The DKNs modify and extend the current statistical and artificial intelligence tools to support the acquisition of data-driven knowledge and the increasing data analysis coming from heterogeneous, structured or semi-structured sources.

The project of complex information systems – in general – and of knowledge nets – in particular – often requires, in order to be useful, a modular project which involves the decomposition of the whole task in more manageable sub-tasks. So, the DKNs are made up of multi-agent systems, composed of more than one agent, more or less autonomous, each one responsible for a given data source (e.g. an autonomously managed database) or characterized by a particular analysis capacity (e.g. knowledge discovery tool).

In order to assure a satisfactory effectiveness of multi-agent systems, the DKNs include some mechanisms for the coordination and control of the set of agents.

In addition to the above-expressed technical problems, it must be taken into consideration few problems concerning reliability, fault tolerance, performances and infrastructures' safety needed to provide the necessary connectivity among the distributed data, the knowledge source and the users.

7.1 Techniques for a distributed knowledge net design

The mobile agents technique, made easier by the recent progress of technology in the communications and in the artificial intelligence field, supplies an interesting picture for the project and for the implementation of communication applications in general, and of the distributed nets in particular.

A mobile agent is an object having a name, containing code, persisting state, data and a set of attributes (e.g. history of the movements, authentication keys) and it is able to transfer itself from one host to another in order to execute the task entrusted to him. The mobile agents provide a potentially efficient picture executing the distributed calculations in sites where the important data are available rather than costly dispatching a bulky amount of data in the net.

Honavar *et al.* [6] planned a DKN made up of the following components:

- a mobile agents infrastructure;
- clever agents for the finding and for the extracting of information in order to obtain a knowledge discovery;
- control and coordination mechanisms for multi-agents systems (MAS).

The system is constituted by modular and objects-oriented extendable tools, to quickly reach both the project and the realization of prototypes of MAS for different applications.

At the moment, research works on the infrastructures for mobile agents (IAM): many IAM projects consist of, at least, three components:

- agents servers;
- agents interfaces;
- agent brokers.

The agents servers support basic mechanisms for the migration, for the authentication and sometimes for other services; the agents brokers supply the addresses of agents server and support mechanisms for agents servers univocal naming.

The agent's interface is used by the application programs to create and interact with the agents.

Many companies and groups of research have recently proposed some standards (MAF) on key aspects concerning mobile agents' infrastructures to make interoperability among different mobile agents easier (different architectures, project and choices of implementation).

The clever agents – software entities which execute specific tasks through users having different autonomy and intelligence levels – offer an attractive approach to the project of DKN nets. Particular interest is given to: *reactive agents* which readily react to the changes they perceive in the environment where they are; *deliberative agents* that plan and act in an objects-oriented mode; *utility-driven agents* that act in order to maximize an utility function built *ad hoc*; agents able to learn that modify their behaviour in function of their experience; and agents who combine different ways of behaviour.

The prototypes of DKN systems implemented by Honavar *et al.* [6] include clever agents used for a smart information finding, for information analysis and for automatic knowledge discovery functions.

The mobile agents adaptable for the information finding need to acquire automatically the user's preferences using machine learning techniques; moreover, they are successfully used for the selective finding of newspapers articles, news, coming from remote sites.

Honavar *et al.* [6] are trying, at the moment, to extend their system in order to make it able to elaborate different data types and information sources for various applications.

The use of data coming from heterogeneous sources that can be found in multiple hardware platforms and operating systems, located in different geographic areas, requires a system which is at the same time univocal, steady and flexible, in order to guarantee the interoperability among several data sources and customers.

The interoperability among different platforms (hardware and software) – if standard relational databases are used – has become more simple to be realized thanks to tools independent from the platform, as JDBC and CORBA; however, the DKNs must be able to provide an access, without additional infrastructures, on data distributed on multiple databases, not linked among them, having got heterogeneous data both in their form and in their content.

It is possible to classify the current methodologies to elaborate data coming from heterogeneous sources into two big categories: *multidatabase systems* [7] that applies traditional techniques to bring up-to-date data contained in the various databases; and the *mediator-based systems* [8] in which new data sources can be added by simply formulating a set of rules describing it.

Honavar DKN project [6] has a pragmatic approach to face the interoperability among heterogeneous data sources. Methodologies coming from both *multidatabase systems* approach and *mediator-based systems* have been used in order to implement an object-oriented data warehouse employing software agents having a knowledge basis.

The specific knowledge concerning the data sources and the application domain has been used to extract data, and to elaborate, analyse and organize them in one or more data warehouses [9] so that both can execute complex query and subsequently analyse the data for future knowledge discovery purposes.

The effort done by Honavar *et al.* [6] consists in inserting different tools in software agents able to manipulate different data typologies (texts and images).

The high amount, the diversity, the variety of data to be analysed in specific applications and the scientifically important but complex relations probably existing between them bring the DKNs to include sophisticated tools for the data-driven discovery of knowledge. For such purpose a variety of automatic learning approaches as artificial neural net, a statistical and syntactic methods, an automatic induction of rules and evolutionary techniques are available.

The nature of knowledge acquisition, the process of discovery and the choice of algorithms or of specific tools to be used to reach the prearranged aim depend on an amount of factors as: general purposes of the knowledge acquisition function (pattern setting, prediction, check); the nature and the amount of the beforehand knowledge of the domain available; the choice of the data (vectors, text, images) and the knowledge representation (decision trees, rules, neural nets); the amount and the quality of the data available for the learning or, if necessary, the knowledge increasing refinement as soon as new data are produced.

8 Application examples

8.1 Mobile agents-based events scheduler

In Ref. [10], Glitho *et al.* presented a "study case" concerning the mobile agents' use for tasks of automatic information retrieval.

The aim is to realize an automatic system of meeting management involving more than one participant. This means taking the following factors into consideration:

- participants' availability
- resources' availability (adequate rooms, overhead projector, etc.)
- participants' preferences
- importance and duration of the meeting.

The problem consists in the identifying of the data and of the space of time in which all the participants are available.

The classical solutions to such a problem are given by electronic organizers shared by means of centralized calendars based on a client–server system whose organization's procedure consists in downloading the calendars of the potential participants in order to identify the time space in which they are available.

The identifying of free time spaces is usually visual; that involves obvious scalability and privacy problems: it is necessary, indeed, that the engagements of all possible participants are made public to the potential organizers of the meeting.

In mobile agents model each participant has a personal agent that can negotiate a potential reorganization on his behalf (fig. 8). The working hypotheses are the following: the videoconference will take place in a certain month and its duration (e.g. one day) is assigned; the conference will be organized only if there are at least four participants.

Acting in such a way the system performances increase and an easy personalization, an independence from the hardware/software, a privacy and automation improvement are obtained.

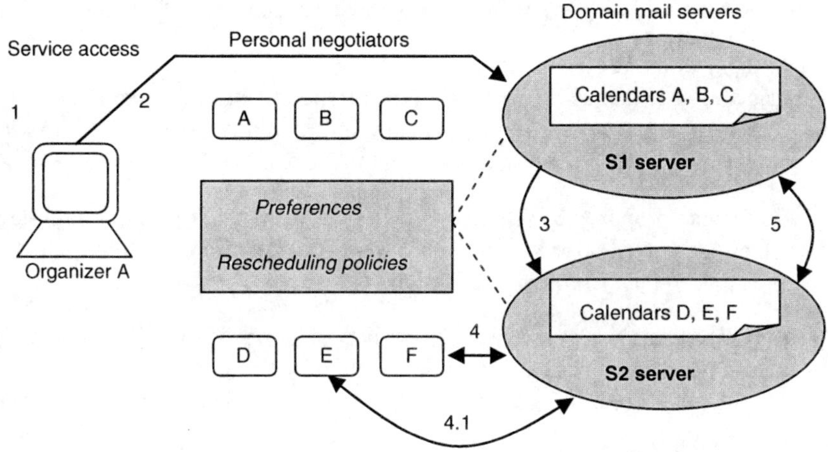

Figure 8: The scenario in which the mobile scheduler is operating.

The above-presented low- and the high-level architecture is described in the Section 8.1.1 [10].

8.1.1 Description of mobile agents' architecture

In the approach presented by Glitho *et al.* [10] a mobile agent is sent to a server in order to retrieve and to elaborate the information at the local level rather than uploading it and elaborating it in a client. The results are therefore conveyable to many applications based on the same principle, as it often happens in information retrieval tasks.

A first observation can be done: the mobile agents' paradigm used to improve the performances of the systems of automatic information finding cannot be necessary; indeed the well-done client–server applications can be sufficient to reach the purpose.

Such an approach, however, can go beyond the application's project and it can require the API extension provided by the server (as shown by the study case [10]); moreover, it cannot be simple to interface mobile agents with well-done client–server application.

A second observation is that applications based on mobile agents can still be better than the client–server approach from the point of view of performances. The mobile agents are indeed suitable for information-retrieving tasks when performances constitute a critical point.

In order to reduce the time of answer in an environment in which there are many calendar servers, many mobile agents are sent in the net; however, the penalization in terms of net's overloading has to be minimized. This factor depends on the number of agents that have to be sent in the net and on the quantity of servers each agent has to visit.

Multipart scheduling events with many mobile agents yields few observations:

- the algorithm must be distributed
- adaptability to the particular problem in question (coordination model)
- investigation on the properties of many information retrieval systems when many cooperating mobile agents are sent in the net.

The system proposed by Glitho *et al.* [10] is divided into two parts: high-level (fig. 9) and low-level (also known as "software architecture", fig. 10).

The prototype is based on four fundamental propositions:

- The calendar server provides an API to access to the information contained in the calendars. Any proposition on the nature of API is given.
- The calendar server has got a platform in order to hold and execute the mobile agents.

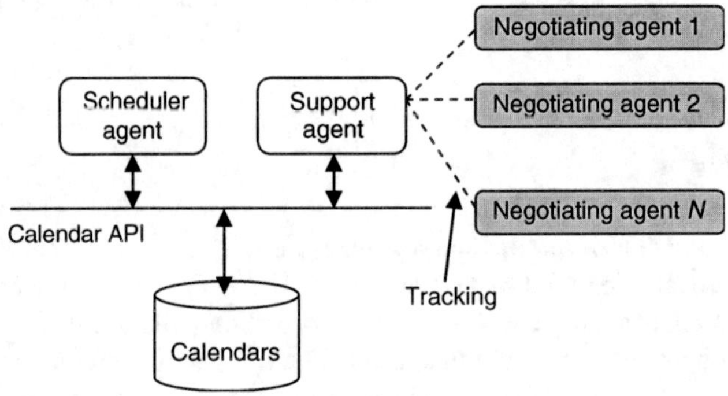

Figure 9: High-level architecture.

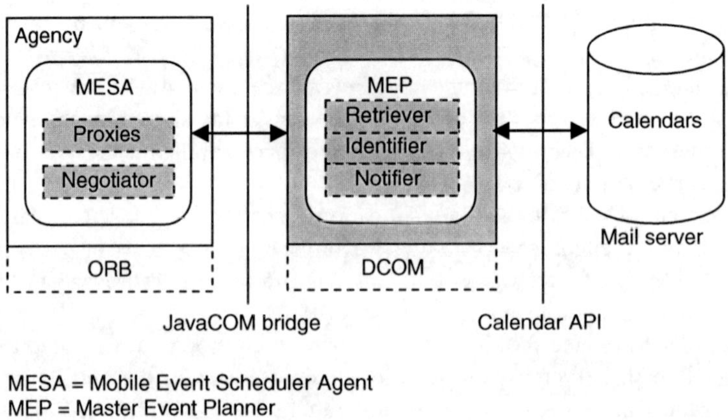

Figure 10: Software architecture.

- The calendar server hosts a support scheduling application having the task to reallocate a meeting previously allocated.
- The application is written in Java.

The main components of the system are: the *scheduler of events*, the *support agent* and the *negotiation agents.*

The *support agent* is a static one and it leaves trace of negotiation agents which can be mobile ones; it acts as intermediary between the scheduler agent and negotiator agents; its task is to reorganize the events and to proceed also in the real reallocation.

The *negotiation mobile agents* can migrate towards the server and remain there during the entire negotiation process. Even though the evaluation of these agents' performances compared to fixed negotiation agents is interesting, the authors preferred to concentrate themselves on the information retrieval aspects.

The scheduler agent's components are

- The *user interface*: it interacts with the organizer of events in order to obtain the list of the participants and the e-mail addresses; it also interacts with each participant to obtain the authorization to access to the electronic organizers.
- The *information retriever*: it interacts with the calendar server in order to find the information requested for identifying the dates in which all the arranged participants are free from appointments.
- The *appointment identifier*: it uses the information in order to derive the possible dates and time slots.
- The *date negotiator*: it interacts with the support agent to negotiate the reallocation.
- The *notification agent*: it sends to the participants the real allocation time.
- The implementation has been limited only to the scheduler agent.
- The mobile agent responsible to allocate the events includes the user interface, an information retriever proxy, a date identifier and a notification proxy. The proxies are used because the APIs allowing the information recovery are not necessarily developed in Java and because the recovered information may not be easily elaborated in Java.
- The master events planner resides on the calendar server: it contains the information retriever and the date identifier. The information retriever is developed in a language that allows an easy access to the APIs supported by the server; the date identifier is also developed in the same programming language; this allows an easy elaboration of the information extracted by the calendars.
- The "bridge" allows the mapping between proxies and the real entities. It should be noticed that the proxies are not needed if it is possible to easily access the APIs offered by the calendar server through a Java program. The software architecture will be equivalent to the high-level one.

8.1.2 Client–server counterpart

The booking of a multi-meeting event is a process which develops itself in several steps. However, some of them are not pertaining to the evaluation of the performances:

- Both in mobile agents and in the client–server approach, the notifications are sent using the same means (e-mail, or MS Outlook planner meeting).
- The authors ignored, at first analysis, the reallocation process.
- Glitho *et al.* [10] have therefore reduced the evaluation in:
- *Information retrieval*: the information is on-line recovered by the mail server, using one RPC rather than a local one on the MEP node, as it happens in the mobile agents approach: the performance in terms of net workload will be very different.
- *Identification of dates*: in the simple client–server approach, all the information necessary to identify the date is recovered and stored in a collection of elements before the identification process starts. In the optimized approach, the information is recovered bit by bit and the algorithm of date identification is applied to one piece at the time. This influences the performances of the system.

For the evaluation of the performances the authors also built up two client–server architectures: the first one, called "simple client server", is based on the same principles of the client server applications which are built in order to evaluate the performances; the second one is instead called "optimized" because it also optimizes the net workload.

8.1.3 Optimization of the client–server application

The authors supposed to have *N* participants and *M* possible days:

- The participant's daily diary is downloaded.
- Once point number 1 is executed for each of *N* participants, in the best of cases (that is all of *N* calendars concerning the diary of only one day are available), the suitable date and the time slot are identified.
- In the worst of cases, all of the *N* calendars will be downloaded for all of *M* days.

Even though the authors chose a daily calendar as discretization interval, the frequency of the events can be changed, each time, according to the situation; the algorithm would remain the same.

Two problems have to be overcome in order to implement the algorithm with MS Outlook API:

- The quite high answering time: the calendar-object must interact with the other elements of its collection before proceeding to find the subset of elements corresponding to the selection criteria.
- The APIs do not allow the connection to the calendars of more than one participant at the same time. That is why it is necessary to be re-connected to the calendar of each participant, each time another daily calendar must be recovered. Unfortunately, the connection to a calendar costs about 17 Kbytes, which brings to a heavy penalization in terms of net overloading.

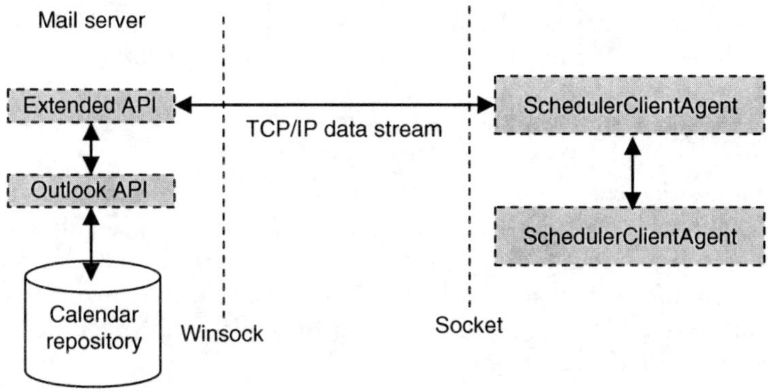

Figure 11: Implementation of the optimized client–server architecture.

Fig. 11 shows the implementation scheme of an optimized client–server system.

The key points are the participants' and the resources' availability. This problem is limited by various constraints: the participants' preferences, the priority and duration of the event, the time slot in which it has to be placed. The general problem is how to identify the date and the time slot in which all the involved participants (or at least a minimum part of them) are available, respecting the constraints shown before.

The problem is complex: the low-priority events can be re-scheduled or even cancelled in order to free space for higher priority events. In Intranet environments, the existing solutions are based on electronic organizers through some client–server centralized systems (MS Outlook). These tools allow the organizers of events to download the calendars of potential participants' in order to identify the time spaces on which the potential participants are available.

They used two metrics: *net workload* and *time of answer*. The net workload measures the amount of data conveyed on the net during the scheduling process. The time of answer measures the process' duration. These two metrics are influenced by three principal factors:

- the N number of the participants
- the I agenda in which each participant is available where $1 \leq I \leq M$
- the density of the agenda of each involved participant

The authors assumed that N can change in a range between 5 and 30 and that $M = 30$; they afterwards put randomly in all the agendas – during the working hours – none, one or two appointments each having a duration of an hour. The duration of the meeting was supposed to change from 1 to 8 h.

The experimental results show that in case of mobile agents platforms the net workload is constant; in the client–server applications the workload linearly grows with the number of the participants, but in the optimized version the slope is lower. The answering time is linearly dependent on the number of the participants, however, when the mobile agents solution is adopted, the slope of the straight line is definitely lower with respect to the other solutions.

8.2 Searching through genetic algorithms

Referring to the clever retrieving framework of information in Internet (fig. 12), it is possible to observe the presence of mobile agents making the research activity starting from a set of input information, seeking for the links of such information and evaluating them through some clever evaluation algorithms as the *fuzzy logic*, mainly used for the searching for web pages similar to the entry ones.

During its life the agent works in many hosts and it gets over the limits borne by the presence of different processors and operating systems: in order to be executed it is not necessary for the agent's code to be preventively installed on every host it has to visit, but it is enough that the host supports the mobile agents.

The mobile agents using genetic algorithms have, as their main purpose, the reduction of the useless information transfer, which yields the decrease of the net traffic.

The "*processor*" is the application responsible to produce and to classify the URLs' database and to manage the interfacing and the execution programs.

It is possible to move to the *user interface*, inside the mobile agent, the software parameters of the genetic and of the evaluation algorithm, as well as the precision rate and the mutation parameters.

In the "*processor*", moreover, a keyword's extent can be implemented through its decomposition in several semantically homogeneous entities. The group of keyword just obtained feeds a suitable database called "*relative keywords database*". In fig. 12, a logical database division is shown: *address database* containing the addresses of the pages to be searched, *searching result database* concerning the results, *relative keyword database* containing the keywords as above mentioned.

The task of the "*user interface*" is to initialize the requests for extraction made by the user and to show the achieved results.

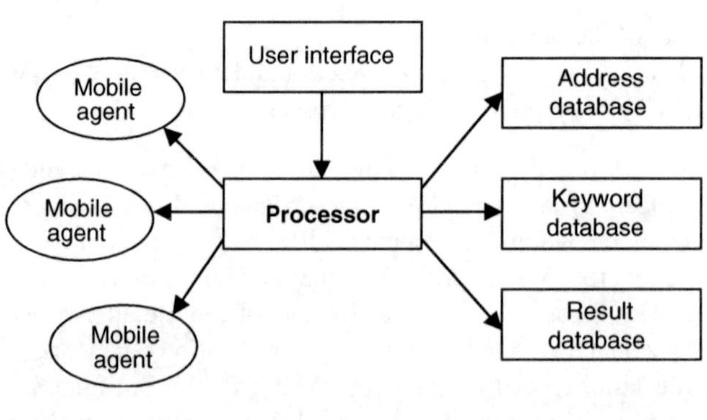

Figure 12: Framework for the information retrieving.

In other words, a smart mobile agent used for the genetic research tasks works as follows:

- Current set initializing: the set of input documents, representing the pool of the current solutions, are first submitted to indexation and then to a process of keywords extraction.
- The documents found through the links of the elements of the current set are compared with the input objects for similitude through the "fuzzy" analysis algorithms, and the best ones among them are inserted in the current set.
- A new set of solutions is produced by genetic operators: a set of URLs is selected by its database in a random way; it is inserted in the current set and the best document among those ones is chosen in order to add it in the output set. All the documents linked by this last one, that is by the element promoted at the new generation, are added in the current set.
- the second and third steps are repeated, till the output set reaches the prearranged dimension or the current set turns out to be empty.

When the documents in the pool of interest contain several links, this approach can be very slow because in order to choose the best elements of the current generation, all the documents belonging to it and the ones pointed by it have to be evaluated.

In the research algorithm presented above, some elements as the quantization of the search quality, the selection of the elements that have to be promoted to the new generation, the representation of URLs string, the crossover and mutation operators assume particular importance.

As regards the URL representation, that is the address of an Internet document; it represents a first coding of the possible solutions for the genetic seeking algorithm. This representation is divided into fields of different length and meaning: the first one is the Internet protocol, the second one is the server's address (net's denomination, server denomination, Internet protocol organization) and the third one provides information about the routing path from the root server to the document at issue.

The crossover and mutation operator allow the passage from the current-generation URLs towards those of the next one: a "gene" is randomly chosen inside a string, to be modified in order to obtain a new "allele", allowing the identifying of the new string and the new URL.

The task of the "fuzzy" analysis algorithm is to count the occurrences of the keywords, and of the words related to them in the document at issue, and to attribute to the document a "weight" in terms of evaluation proportional to such occurrences. Among the numerous analysis algorithms, this one is characterized by a low time of evaluation.

In the mobile agents platform used for the clever search presented above, the mobile agents are sent in the sites where the useful documents are stored and where they implement the evaluations taking back only the results. The genetic evaluation algorithms apply the "temporal locality" and "spatial locality"

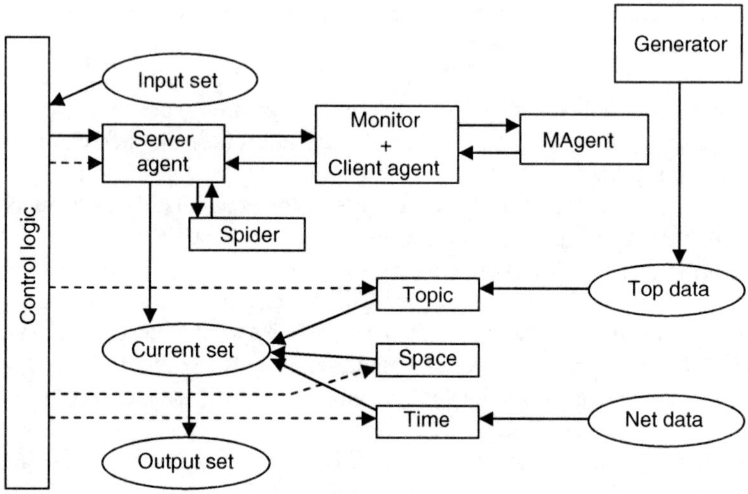

Figure 13: Block diagram of a possible implementation.

principles. This last one means that all the explorations are implemented in the environments close to server where the father-document is located such as the same server or the local network; the "temporal locality" instead refers to the conservation of the elements according to the foreseen results and to the application to a subset of such elements of the mutation operator.

The mobile agents are sent to more than one site at the same time; they execute the documents evaluation in parallel on remote servers and only the results are sent to the home server. The technique just explained brings good results in terms of times of return, memory utilization and network traffic.

A scheme of a possible implementation is presented in fig. 13, where the continuous lines show the data flow, the dashed ones indicate the control flow, the rectangles identify the applications and the ellipses show the input and the output data.

In particular blocks are distinguished in:

- "*Server Agent*" is the application executed in the local server for the coordination of several MAgents placed in the remote servers, and for signalling the best documents suitable for the output.
- "*MAgent*" is the application having the evaluation algorithms sent to the remote server, which will feed back only the results of such evaluation.
- "*Monitor + Client Agent*" allows the communication between the Server Agent resident in the local host and the several MAgents resident in the remote sites. In particular, if the used platform is the Concordia it is possible to use some of RMI (remote methods invocation) to keep the Server Agent and the MAgent informed about the messages delivering.

- "*Topic*" is the application that implements the mutation; its task is to select the URLs by the previously created database and to insert these ones in the current set.
- "*Generator*" is the application producing the URLs database. This database is used for the mutation and each of its elements is constituted by two fields: URL and topic. The input parameters of the applications are stored in this field.
- "*Spider*" is the application used to seek for the documents in the "best ones" set received from the Server Agent. The documents are stored inside the local disc: a new folder is created for each remote server containing hyperlinks to the documents on the local disc.
- "*Space*" is the application responsible for the mutation performances; it uses the spatial locality principles. If a genetic algorithm looks for an element having a high value of adaptability on a determinate site, it is possible that some similar documents are located there or in the "close" network. In a circumscribed area of the network it is therefore possible to find the addresses of the documents concerning our seeking.
- "*Time*" is an application referring to the temporal locality principle that takes part in the mutation process; it has a database containing the URLs coming from the output database and it also has a variable whose value refers on the number of times each URL is present in a given set. The mutation process is implemented through the insertion in the new-generation output database, of the URLs with a higher counter.
- "*Control Logic*" coordinates the static and dynamic applications of mobile agents system. It is responsible for the management of database described above: it looks for every element of the output database in the NetData; if the search has a positive result the counter is increased by one unit; vice versa the element is inserted in the NetData after having set its counter value equal to one; in the overflow case, that is when the NetData touches its defined maximum dimension, the elements having the counters with lower values are cancelled in order to create new free space.

To evaluate the mobile agents' positive contribution in a genetic research algorithm it is needed to consider the static implementation of the same project presented in the initial figure, where the "control logic" has been replaced by the "control program" that keeps only the tasks linked to the static components of the first one. The "spider" activities also become different; in fact it does not look for the "best documents" for the output (as we said in the previous case) any more, but it also implements a wide ray searching in Internet (fig. 14).

In this way the overhead rate is high because all the documents are taken to the home server, off-line examined and evaluated, so that enormous memory space is taken up.

In the mobile agents system, the evaluations are executed in the server where the documents are stored, and only the "MAgent clones" can go through the net; this yields a decrease of the amount of the memory used and of the net traffic. The net result is a decrease in run-time and an increase in an information transfer quality.

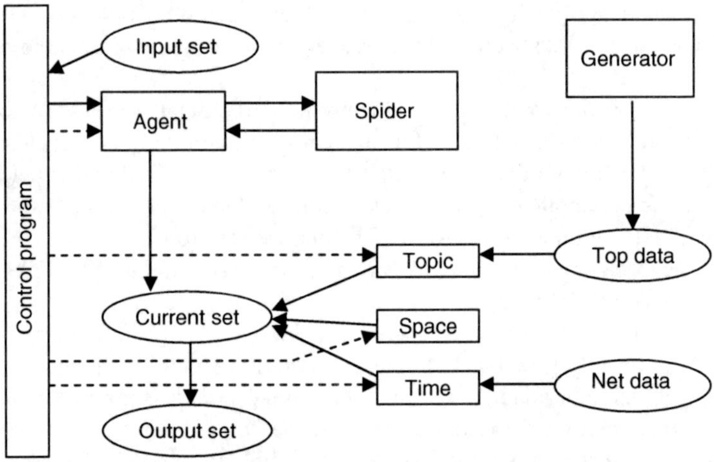

Figure 14: Block diagram of a static implementation.

It is furthermore observable that the "time of return" is in inverse relation both to the search paralleling process and to evaluation degree.

Some graphics summarizing the above description are shown later.

8.3 Smart system

The mobile agents are commonly used in information retrieval applications. The choice of this paradigm allows the applications interacting with it to avoid the transfer of intermediate data, to continue the finding task even if the connection with the client stops, and afterwards it allows the merging and the filtering of the results obtained by the single collections of scattered documents inside the net. All this can be obtained sending the agents to the same information sources and elaborating the information at a local level, instead of sending all data to only one central computer where the elaborations will be executed.

Since many IR applications simply need to invoke – in a sequential way – a series of operations requiring a modest amount of code in order to decide what operation should afterwards be invoked, and since such applications are linked by the time of execution of the operations on the server, it is possible to affirm that a mobile agent can have good performances even when implemented in one of the interpreted languages used in many mobile agents systems.

The smart system is a statistical system of information retrieval that uses the vector-space model to measure the similitude between documents. It is placed inside a stationary agent that provides a multipurpose interface that:

- executes one textual query and obtains a list of important documents;
- obtains the entire text of a document;
- obtains a similitude scores for each pair of documents in a list of documents.

Such similitude scores are used to build different graphical representations of the results of the queries.

As soon as an agent begins the execution, it registers itself on the yellow pages which are a simple distributed directory service, organized in a hierarchical way. When the registration is completed, the agent register supplies its location and a set of keywords describing its service; at the same time it sends a query to the yellow pages which return, as feedback, the location of all the services whose keywords coincide with the sent query.

The agent's principal application is a GUI that is executed inside the user's computer: the user inserts a free text query in the GUI and it optionally selects specific collections of interesting documents from a list of known ones. Once the GUI has the query, it sends a mobile agent inside the local PC that consults one or more local agents specialized in "hearing" the net and in keeping trace about the connections between the user's PC and the rest of the net. These agents know the type of net hardware present in the machine, the maximum hardware band, the history of the up/down times of the net link, the latency observed at the moment and the net link band.

The timing uptime/downtime is used to calculate the reliability factor, which is the probability that the net connection suddenly collapses in the following minutes.

After having consulted the net agents, the system takes the most important decision. If the net connection between the user machine and the net is reliable and this last one has a wide band, the agent is in the user's location; otherwise, if the connection is unreliable or the net has a low band, the agent jumps in a proxy site inside the net.

The proxy site is usually chosen by the agent in a dynamic way; otherwise this last one would migrate at a proxy site, consult the yellow pages to determine the location of the documents collections (assuming that the user does not select specific documents collections) and interact with the stationary agents used as interface. Here, the agent takes his second decision: if the query requires only few operations for each documents collection, the agent simply invoke some RPC calls through the net; but if the query requires many operations for each document collection, or if the operations involve a high amount of intermediate data, the agent sends out some child-agents that travel towards the documents collections and implement the query operations locally, avoiding the transfer of intermediate data.

When the principal agent receives the results from each child-agents, it melts and filters those results, goes back to the user PC with the list of the final documents and shows it through a GUI. Even though the behaviour shown by this agent is complex, it is verily quite simple to be implemented and involves only 50 tcl code lines. In particular, the decision of using a proxy site and of creating child-agents involves a few "if" statements that check the information returned by the net sensors and by the yellow pages. It is difficult to image that any other technique would equally allow a flexible solution with the same (little) amount of work.

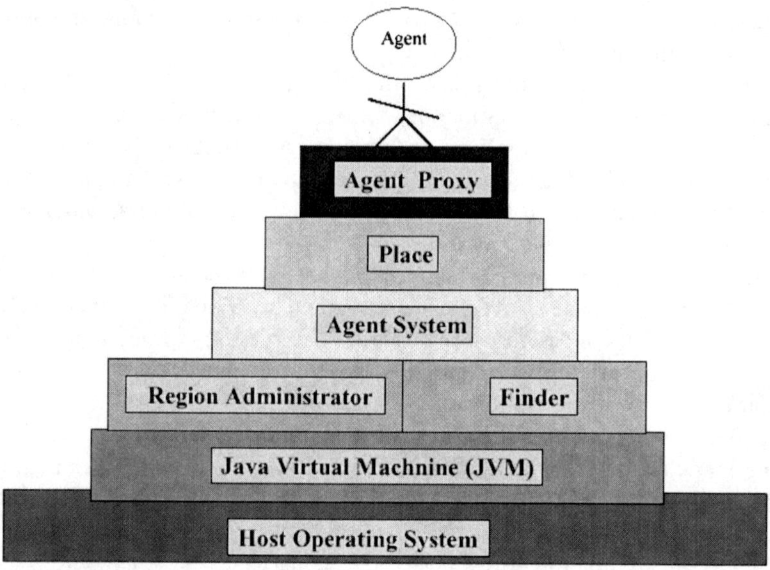

Figure 15: The SMART layered architecture.

The critical inefficiencies involve the communication and migration overloading. Because of the communication excess, the agent behaves worse than the client–server approach if it chooses to remain stationary. Likewise, because of the migration overload, it is better for the agent to migrate only when the conditions of the net are poor, or when each query requires a high amount of operations for each collection. Fig. 15 shows the SMART architecture.

8.4 JAM

Stolfo *et al.* observed in Ref. [11] that it is not possible to apply the algorithms of automatic learning directly to databases of big dimensions since the time of answering would be prohibitive; furthermore, most of the algorithms existent in literature require that all the data must be physically in memory to execute the elaboration.

Starting from such considerations, the authors introduced the JAM system (Java agents for meta-learning over distributed databases) that uses a general approach to scale the data mining algorithms, called by authors "meta-learning".

JAM provides a set of learning programs, implemented in Java to calculate some models starting from data locally stored in a site. The JAM architecture is a system based on agents that calculates meta-classifier on distributed data.

JAM also supplies a set of "meta-learning" agents and a distribution mechanism in order to combine and to allow the migration of some models previously calculated. One of the JAM applications was the identification of the intrusions in net and the discovery of fraudulent actions.

8.4.1 Meta-learning

The meta-learning supplies a uniform and scalable solution improving the efficiency and the accuracy of inductive learning when applied to high amount of data in nets of wide dimension and moreover offers a span of several applications.

The approach suggested in Ref. [11] is to execute in parallel a certain number of learning processes, each of them implemented as a classical serial program, on a certain number of data subsets, using therefore a technique of data reduction. At the end, it will be necessary to combine – through the meta-learning – the several obtained results. This approach has two advantages:

- it uses the same serial code in more than one site without the difficulty to write parallel programs;
- the learning process uses little data subsets that can therefore reside in the central memory.

The accuracy of the concepts learned by several separated learning processes can be lower than the serial version based on the only one process applied to the whole database, since most of the information cannot be accessible to each of the independent and separate processes.

On the other hand, combining these concepts of higher level through the meta-learning can improve the precision levels, in a way comparable to the ones obtained with the serial version applied to the whole database.

This approach can furthermore use a variety of different learning algorithms on several platforms.

Because of the proliferation of the workstation nets and of the growing number of new learning algorithms, the approach presented in Ref. [11] is not referred to a specific parallel or distributed architecture, nor to a particular algorithm, so the distributed meta-learning can suit itself to new algorithms and to new systems: in such sense the approach is scalable, portable and extensible.

8.4.2 JAM architecture

It is a system based on agents and planned as an operating systems extension. It is a system of meta-learning supporting the diffusion, in the net, of learning and meta-learning agents towards distributed databases.

JAM is implemented as a collection of distributed learning and classification programs linked among them by a "*datasite*" net.

Each JAM datasite contains:

- a local database;
- one or more learning agents, or in other words, programs able to migrate towards other sites as applet Java, or to be stored as native applications callable by Java applets;
- one or more meta-learning agents;
- a local user configuration;
- services of animation and graphic interface.

The JAM datasites have been planned to mutually collaborate in order to exchange classification agents, calculated by the learning agents.

The local learning agents work on the local databases and calculate the classifier of the local datasite. Therefore, each datasite can import the remote classifiers from another one and combine it with its own classifier, using a meta-learning agent: at last, after the meta-classifiers structure has been calculated, the JAM system manages the execution of all modules to classify and label the datasite of interest. These actions can take place concurrently and independently in all the datasites.

The datasite owner manages the local activities through the local user configuration file. Through this file it is possible to attain the parameters needed to execute the learning and meta-learning tasks. These parameters include the names of databases to be used, the policy of subdivision of the training and verification subsets, the local agents that have to be sent, etc.

Besides the static specification of the local parameters, the user can use the graphic interface to supervise the exchanges among the agents and dynamically manage the meta-learning process. Thanks to this graphic interface the owner can access to several information as the accuracy, the trend statistics and the log files, to compare and to analyse the results in order to improve the performances.

The configuration of a distributed system is pertaining to the configuration file manager (CFM): a central and independent module responsible for keeping up to date the system's state. The CFM is a server providing information about the participating datasite and about the log files eventually consulted for references and future evaluations.

The Java technology has been used to build up the system's infrastructure, to develop specific operator-agents which make up new agents and implement a GUI.

8.4.3 CFM (Configuration file manager)

The CFM fulfils a task like that of a DNS of a net. It supplies the registration services of all the sites that wish to become members and to participate in the meta-learning activity.

When the CFM receives a *Join* request from a new site, it verifies both the validity of the request and the datasite identity. If such procedure has success, the CFM accepts the request and sets the site's state as *active*. Similarly, the CFM manages the request for "*start*": it sets the site's state as *inactive* and removes it from the set of members. The CFM keeps the list of active datasite's members in order to establish the contact and the cooperation among them. Apart from that, the CFM keeps the information concerning the groups that have been built (which *datasites* collaborate with which *datasite*); it registers all the events and shows the state of the system. Through the CFM, the JAM system administrator can analyse the participating datasite.

8.4.4 Datasites

While the CFM provides a passive maintenance function, the datasite constitutes the active members of meta-learning system. They in fact manage the local databases, obtain the remote classifiers, build the local base and the meta-classifiers and – at last – interact with the JAM user.

Datasites are implemented as Java multithread programs having a parti cular GUI.

After the initialization, a *datasite* initializes the GUI, through which it is able to accept data input, show the state and the results, register with the CFM, instantiate the local learning agent and create a socket server in order to "listen" the connection from the correspondents datasites. Therefore, it waits for the next event through an open socket.

In both cases the *datasite* verifies that the input is valid and that it can be served. Once it has been established, the *datasite* allocates a separate thread and executes the requested task. This task can be one of the JAM functions: calculates a local classifier, begins a meta-learning process, sends local classifiers to the corresponding datasite or asks them for remote classifiers, takes back the current state or presents the calculation results.

8.4.5 Classifier

JAM supplies visual tools helping the users to understand the learned knowledge. There are several types of classifiers (decision trees with ID3) that can be represented by graphs. In JAM there have been adopted the principal components of JavaDot, that is an extensible visualization system that allows the user to analyse the graph.

Since each automatic learning algorithm has its format to represent the learned classifier, JAM uses a specific translator to read the classifier and produce a JavaDot representation.

8.4.6 Animation

For demonstrative and didactic purposes the learning component, that is the JAM user interface, contains a collection of animation panels that visually show the learning phases at the same time of execution.

When the animation is established, a transition in a new calculation or analysis phase identifies the beginning of the animation sequence corresponding to the below activity. The animation keeps on running until the activity stops.

The JAM program also provides the user with the choice to manually start each learning phase or to run the process in automatic execution.

8.4.7 Agents

The scalable JAM architecture allows to insert with easiness other learning agents designed as objects. JAM provides the definition of agent's parent-class and each agent instance (e.g. a program implementing a given learning algorithm) is therefore defined as a subclass of this parent-class.

Among the definitions that are inherited by all the agents' subclasses, the parent-class provides a simple and minimal interface under which all the subclasses must suit. If an agent conforms itself to this interface, it can be put in and immediately used inside the JAM system.

In details, in JAM an agent needs to have implemented the following methods:

- A constructor method. Without arguments for this JAM can instantiate the given agent that knows its name. This constructor can be supplied by the data-site owner through the local configuration file or the graphic interface.
- An initializing method. In most cases the agents' subclass inherits this method from the agent's parent-class. Through this method, JAM is able to provide the arguments needed by the agent. The arguments include the name of training and verification database, the dictionary name and the output classification name.
- a *buildclassifier*() method: JAM invokes this method to tell the agent to learn or meta-learn by a given training set.
- the *getclassifier*() and *getcopyofclassifier*() method: These methods are used by JAM to obtain the last versions of classifiers that can be inserted in any datasite involved in the net in question.

It is to point out that JAM architecture is designed and implemented independently by the learning algorithms of interest. Furthermore, it is extremely simple to insert any learning agent satisfying a minimum of interface requirements. This characteristic makes JAM an extremely powerful and extensible product for data mining services.

8.5 Information filtering

The selection of the information has been applied in most of the various domains; an example of application is the "*Agent Know-How*": an e-mail-agent used in the social navigation field and utilizing the e-mail systems and the real knowledge of social nets. It implements the retrieving and the filtering of the information through the term frequency/inverse document frequency (TFIDF) method and a similitude form through the examination of the documents that arrive little by little in order to find the most suitable person to whom to send the request for the user.

Yenta is a multi-agent system connecting people depending on the interests and on the activities. Yenta agents are grouped in such a way to represent the user's interests and to define the specific users' profiles; they build a similitude matrix among each characteristic, both locally and remote.

Junk-e-mail is a filtering system acting in such a way to delete not-wished and not-requested messages: it uses some Bayesian models, supplemented by particular characteristics to improve the knowledge and the discriminating ability of the system.

8.6 Identifying and information discovery

Amalthea is a multi-agent system that finds out and collects information from several sources of users' interest and shows the searching results in a synthetic form. It is based on two general kinds of agents: those for the filtering of the information, responsible for the elaboration of the custom-made profiles of the users, and those for the information discovery responsible for finding and downloading the information requested by an user. Discovery agents act as meta-models of search engine; sometimes – depending on the cases – they check a previously visited site and analyse the document to find out if it is changed and how much it has changed through the use of a database where the URLs are stored.

Do-I-Care is a multi-agent system helping the users to find out the changes in the web through the use of technical and social tools. Each agent checks web pages previously found important for few changes; if there are modifications in the interested pages, the agent notices all the changes. The primary activity of the *Do-I-Care* is to provide aid to the user rather than to find out new pages.

8.7 Spider or indexes systems

Web Hunter is a web robot based on keywords and having a few collaborative filtering aspects. It starts from the user's location and looks for all the link or URL correlated to the original one, according to the user's target (single words or sentences).

Info Spiders is a multi-agent system suitable for web pages on-line searching. It is based on a technology designed to recover the URL where to obtain important documents for a given query. *Info Spiders*, as *Web Hunter*, uses a basic set of parameters to begin the execution.

8.8 Aided navigation systems

Letizia is an application based on agents having a user interface. The agent outlines the user's behaviour and tries to obtain results of interest through a concurrent and autonomous link exploration from the user's current position. "*Let us browse*", as *Letizia*, is a web browsing agent, helping a group of people in the navigation and providing a requested information of common interest. Its aim is to explore the net, to find the common interests, to stimulate the interaction and to use the interests of groups of people so that the information recovered by the web can be filtered.

"*Margin notes*" is a "remembering agent" having many aspects concerning the information's discovery field. It actively helps the user, observes his context and suggests information related to his interests. It works in background without the user's intervention, providing little information from time to time, in such a way that it does not bother the user.

8.9 Mobile information extractor

Hae Kong *et al.* [12] developed a mobile information extractor based on mobile agents in order to provide various information for the IHWA (Information Harvest Warehouse) information supply system, whose information is relegated at the affiliated sites.

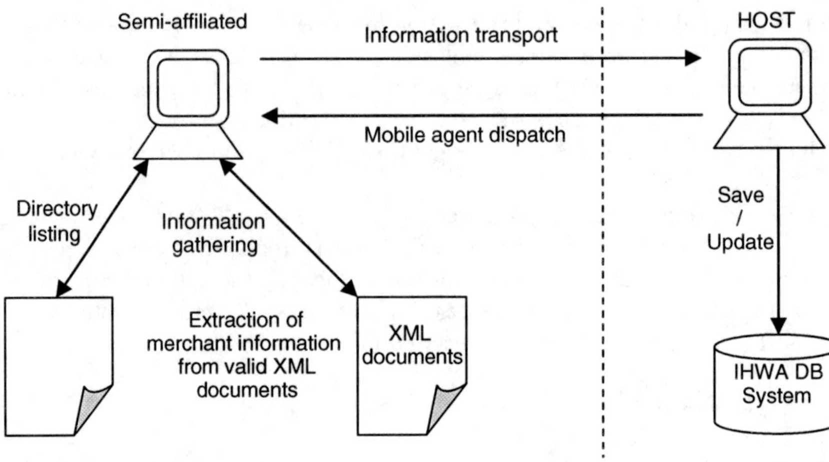

Figure 16: A mobile information collection system.

A mobile "collecting" agent has been developed through aglet that is a mobile agents environment designed by IBM (fig. 16); it has been assumed that the target documents situated in remote sites are drawn up in XML with their DTD.

The following operations have been implemented in detail: the mobile extractor has been created; it sends the agent to remote and not IHWA-affiliated sites, looks for documents through the directory listing, analyses the XML documents, extracts the necessary information and transmits the SQL information back towards the host. Therefore, an agent resident in the host receives the SQL messages and stores them in the IHWA databases. At last the sent agent goes back to its host and it is deleted.

The extractor identifies the necessary information from the remote sites which are not affiliated and have a few restrictions concerning their documents format. To improve the capacity of the supplying information of the IHWA system, the mobile agent goes to a site, where it examines the XML remote documents, extracts the necessary information, sends it back and stores it in the IHWA database.

The authors built a system to extract the proper information from XML documents using some mobile agents' operations previously shown. The extractor agent has been built to identify the directory listing, XML parsing, and to retrieve information by remote sites.

A host and a remote agent are created; later the last one is sent towards a site. The remote agent examines the directory through *Hparser* in order to look for XML files. XML parser finds both the elements and the attribute of the XML document through the use of *SAX* [12]; therefore, the remote agent creates an SQL listing that is sent back towards the host agent. At last the host agent stores the SQL in the IHWA database.

Once all the information's collection activities have come to an end and the gathered information has been sent and saved in the host, the agents are removed.

The advantage of using this mobile agents system is to allow us to provide a more detailed information to the users. It is furthermore possible to considerably reduce the net traffic and the host elaboration overload.

The pseudo-code of the mobile information extractor is shown below:

Host side:

- agent creation
- remote agent creation
- events listener setting
- host agent setting
- obtaining a remote site's URL
- sending the agent towards a remote site
- application of SQL transmitted to the IHWA databases
- agent deallocation

Remote site side:

- information extractor creation
- directory listing (Hparser)
- XML (SAX) documents parsing
- information extraction
- SQLs creation
- Transmission of SQL messages to the host
- agent deallocation

Thanks to such technology, this information extraction system is able to provide to the customers some richer information from both affiliated and non-affiliated sites.

Moreover, since the elaboration takes place in remote and only the minimum information is transmitted, the mobile information also minimizes both the elaboration time of the documents by the host and the transmission overload on the net.

Such efficiency can solve the problem of the documents elaboration presented by spider-based conventional methods. In order to implement high-quality information extractors, it is possible to plan the mobile agent optimization to make it able to work dependently from the site.

8.10 Multi-agents platforms

In the last decade, main AI's interests have been focused on knowledge-based distributed systems. More precisely, it is only lately that the MAS have produced remarkable progress. These systems have been realized by autonomous and cooperating agents, each of them playing a specific role in the company and supplying the planner an organization for heterogeneous systems' interoperability.

The main characteristics of such systems are:

- *modular structure*: MAS-based system is a construction made by basic blocks that cover all the hardware/software subsystems surrounding the whole system.
- *encapsulation*: the software components are encapsulated by a common interface, overcoming the problem of the heterogeneity.
- *cooperation*: a mechanism for the clever cooperation should be supported in order to allow complex interactions among the framework components.
- *distribution*: the location of the modules should not be forced. The system should be able to distribute them according to any request.
- *opening*: the cooperation among heterogeneous components must be also understood as cooperation among heterogeneous systems in order to allow heterogeneous MAS to cooperate.
- *easiness of use*: CORBA-based systems are neither easy nor comfortable to be used; they are complex and they requires a high initial effort by the programmer. MAS incorporates some powerful abstraction mechanisms and easiness in development.

Gonzalez *et al.* [13] developed MAST (multi-agent system tool) for the working out of distributed systems in an agents-oriented environment. It provides mechanisms to select the agents individually, specifying the offered and required services and particular goals. It incorporates also the agents' groups definitions, tools for the representation and the exchange of knowledge, and the coordination mechanisms among agents.

8.10.1 Mix architecture

Mix is the reference architecture for all the systems developed with MAST. It presents two fundamental models: agent model and net model.

The first one defines an agent as a set of elements:

- *Services*: functionalities offered to other agents.
- *Aims*: functionalities that an agent takes ahead for its interest (not as result of a petition among several agents).
- *Resources*: information about outside resources as services, libraries, ontologies, etc.
- *Internal objects*: data structures shared by all the processes that can be run by the agent to take requests for service or to achieve aims.
- *Control*: specification about how the requests for service are manipulated by agents.

The net model defines two different types of agents: the *net agents* providing net management services and the *applicatory agents* supplying their own services according to the particular role they have in the system.

Such model is structured in three levels: interface, message and transport layer.

The first one offers an API with C++ and Java and supplies communication services among agents through messages exchange. The second level offers services for the management of the addresses and the purposes of the message. The transport layer offers basic functionality to send and receive messages through the TCP/IP protocol.

8.10.2 Agents' definition in MAST

ADL (agent definition language) is a declarative language supplemented in MAST to define the agent's outside, while the agents' inside depends on the particular implementation. Two agents sharing the same ADL definition could be implemented differently and even in different programming languages. ADL allows to define the agents in a hierarchical way using well-known concepts of class and inheritance. Moreover, the supplied services, the aims to be achieved, the necessary resources, data internal structure and the possible policies to be applied can be specified in ADL as it is possible to see from Ref. [13].

8.10.3 Exchange information objects in MAST

Once the agents surrounding the systems have been defined, there is the need to implement the knowledge exchange among them. CKRL (common knowledge representation language) is the language used in MAST to declare the knowledge objects which are to be sent, received and managed by the agents.

MAST is used as a developing platform, and MASCommonKads as methodology to plan MAS. Architecturally, generic data mining system defines the following roles held by the agents:

- *User agent*: it collects all the parameters from the interface surrounding the data mining request for service; it requires and waits for the results which have to be opportunely shown to the user. Such agent should provide services to manage the results depending on the system's topology in order to help user to understand them.
- *Group of agents providing denomination services*: some agents, able to locate suitable agents for automatic learning tasks, have to take place in the system, but a high number of automatic learning algorithms have to be potentially supplemented on the systems. Both the previous reasons naturally yield to the construction of a group of agents providing denomination services in order to assure the learning agents' extensibility to other machines.
- *Negotiation agent*: once that all the available services of automatic learning for the request of the user are located, such agent starts a negotiation process with all the learning agents asking them to decide which of them will serve the query.
- *Automatic learning agents*: each of such agents will encapsulate a learning algorithm with its dynamic characteristics describing the learning service supplied in that moment.

- *Control agent*: once the negotiation phase has finished and the automatic learning algorithm has been chosen, the control agent will plan the task to coordinate and check the whole automatic learning process.

A possible scenario is the following: the user agent executes a request for service *S* to the negotiation agent. Such agent bounces the request to a suitable denomination service agent. The denomination agent looks over the possibility to meet the request; if it cannot it delivers it to another denomination agent and waits for the results. The denomination agent receives, as answer, that the agents A, B and C are suitable for developing the requested task. This information is fed back to the negotiation agent which takes a decision on what agent must be chosen. After the choice has been made, the denomination agent informs the control agent which plans the tasks and begins the learning process, sending the service primitive functions to the chosen learning agent. After the chosen learning process has been concluded, the results are sent back to negotiation agent who delivers them to the user agent which realizes a feedback to its decision process in order to improve the meta-learning.

8.10.4 Meta-learning

If the data mining process is observed from a high abstraction level's point of view it is possible to see two parallel tasks: the elaboration and the recovery of the user request. If these tasks are orthogonal the metadata supply the general and the preliminary measures on the data source. The user's aim is to inform about his intention while the QoS one about his preferences. The negotiation process recovers these three parameters in order to take a decision on which – among the automatic learning algorithm – is the most suitable, and starts to control it. After the decision has been taken, the learning process can start producing a data model for the user, giving back more accuracy to the negotiation process in order to improve the meta-learning one.

Metadata: it can be useful to have a "summary" of the data sources in order to take decisions about the best algorithm to choose and to be used. In the *StatLog* project [14] several measures are considered: simple, statistical and correlated to the entropy.

The simple measures are identified immediately: number of classes, attributes and present samples.

The statistical ones have a high computational cost which makes them inadequate because of the dimension of treated data. Other group measures based on the entropy have turned up to be interesting.

- *User's aim*: they are typical description (group, classification, syntheses, subordination models) and prediction tasks (regression, changes identifying, deviations identifying)
- *QoS*: it can be defined by the following parameters:
- *Accuracy*: the wished quality of the new knowledge can be whether interesting or not for the user. He has the chance to ask for the accuracy in the fuzzy way.

- *Time restrictions*: the interactivity of the learning process is observed by the system through this parameter that is correlated to the previous one.
- *Results morphology*: the user can declare his preferences through decision trees, visual maps, sets of rules, histograms, etc.

After having obtained both these parameters and a negotiation mechanism based on clever agents that informed us on what is the best learning technique, the learning process and a control mechanism can start together to carry out the learning task.

Once the results are produced, a feedback is provided to the negotiation process through the meta-learning in order to improve the process of decision.

Three different approaches have been observed for the choice of the best data mining algorithm to be applied in each session:

- the first one, in order to build a kind of expert system, works well with off-line decision rules based on its past experience and on the result obtained by others;
- the second one concerns the learning obtained through the observation of the user's preferences about the algorithms: this approach works well in a system in which the user, if possible an expert one, has the total control; in this way the meta-learning process has its behaviour based on the user.
- the third one is based on the learning through support/reinforcement: the user rewards or punishes the decision system.

8.11 Clever mobile agents to classify documents

Yang *et al.* [15] merged a TFIDF classifier in a mobile agents system on the Voyager platform. In such system, as first step, a mobile agent has been created having the task to look for a set of documents and recover it from a remote site satisfying a user's given query.

The agent is sent to the remote site and recovers the documents which satisfy the query, delivering them to the local site; soon after having executed this task, the agent is destroyed.

The user receives the documents satisfying the query and labels them as "interesting" or "not interesting", that leads to the birth of a data set which can be used to train a "*classification and recovery agent*" through the use of an automatic learning methodology created *ad hoc*.

The "*classification and recovery agent*" based on the TFIDF methodology, planned using the training data, is therefore sent to a remote site in order to recover those documents which have been declared as "interesting" by the classifier. These documents, therefore, are sent to the local site. Once the operation is finished, the agent is killed.

Yang *et al.* [15] tested the approach stated above through a set of variable mobile agents in order to classify an amount of documents, articles or abstracts.

A query is initially entrusted to a mobile agent taking as result the recovery of a document collection that is then used to refine the agent's behaviour through the training documents' labelling provided by the user.

Once the classifier has been suited, it characterizes a mobile agent whose task consists in selectively recovering the documents from remote collections.

It has been discovered that the TFIDF approach [15, 16], used to design information finding agents, has worked quite well in most cases assigning the correct classification close to 90% of the documents which had not been used for the classifier's training.

The advantage obtained through the use of mobile agents' paradigm consists in the agents working in remote sites and recovering only a subset of important documents which are then sent to the local site, rather than unloading all the documents from the distributed databases.

In the experiments led from Yang *et al.* the amount of data moved by the mobile agents (the classifier itself and the obtained important information) appeared lower than the one that would have been obtained using a client–server conventional system. The authors showed that the low traffic produced on the net connection, through the use of mobile agents, is favourable even if high amounts of data have been treated.

Some interesting developments shown by the authors range over – but they do not run out – the following fields:

- *design and implementation of multi-agent systems* in which the agents collaborate to recover and analyse data from databases and distributed knowledge sources, providing a decisions' support in real applications;
- *systematically studies* on performance concerning the advantages and disadvantages of alternative designs of such systems (static and mobile agents' hybrid systems included);
- *design of information analysers* suitable for a wide range of structured and semi-structured data sources.

8.12 Applications suppliers

The ASP paradigm has recently emerged for the management of the applicatory software in the middle-sized companies. The pivoting principle on which ASP is based is that of "renting" the software, that is, the companies register themselves in an ASP and use the applicatory packages provided by the ASP paying only for their utilization, rather than buying the software and installing it. Such a paradigm is especially useful for the little and middle-sized companies in order to knock down the costs of the software.

The emerging of ASP is strictly correlated to the e-commerce. Technologies like e-commerce provide the chance to compete in the world market, which in the past was only under predominance of the big companies.

DDM can be a service supplied by an ASP in an electronic commerce environment.

When the DDM moves from the borders of an organization to become a generic service provided by an ASP and it is used by different entities in an e-commerce system, some further requests have to be satisfied, such as the users' invoicing based on the estimated costs and on the times of answering, to obtain a general improvement of the performances used to manage the data in real time, to be able to fit the various data mining demands with the different client organizations.

If an on-line electronic trade centre is considered, which consists in customers, vendors and one trader, the customers enter into the centre through a web interface and they interact with the vendors through the trader.

At a first level, the trader supplies the services' catalogue to the customers in terms of vendors' profiles and availability of objects and services. At a second level, the trader negotiates the transactions between the customers and vendors. The necessity of a DDM system in such a background rises from two possible sources: the vendor and the trader.

The data mining's requests of the vendor have their origins in the traditional data mining's applications such as the market basket analysis. The necessities of the trader's data mining will be focused on the customer's profile in order to improve the level of service provided to the single customer. Since the environment is intrinsically distributed and heterogeneous, the attention is centred on DDM. In addition to the problems of distributed data, a further complexity due by the optimization of time of answering has been added by the e-commerce.

For instance, when a customer asks for a product which is not available at present, the trader could provide the customer some details such as information about the availability of the product through the analysis of an archive of the past transactions or the availability of similar products offered by the vendors. The trader could also encourage the customer to await on the basis of possible correlations with seasonal offers.

Traditionally, both the trader and the vendor have their data mining systems to manage their business needs. However, the emerging trend of ASP is to supply a means for a generic DDM service. The advantage obtained is that such an approach allows the organizations to access to the data mining services without being worried about the initial costs. In addition, such a service would have the advantage to be quite flexible to incorporate a set of data mining algorithms which the ASP and/or different users can offer as supplemented service.

A framework holding the role of DDM inside an e-commerce background is composed of:

- *customers*: they are people using on-line sale centres in order to buy objects and services.
- *e-commerce systems*: they supplies the infrastructure for the on-line sale centre; it includes a web interface, an electronic catalogue, an intermediary and a database. The web interface is the access point for the customers in the sale centre. The electronic catalogue is the directory of the objects and the offered services, plus the vendor's profile. The intermediary negotiates among the customers and

vendors. The database is used to keep the details about the transactions, the information about the vendor and about the customer, all information collected in order to be used by the electronic catalogue and by the intermediary.

- *vendors*: they are the businessmen who use the on-line sale centre for marketing and selling their products.
- *ASP*: it supplies the application services to the members of the e-commerce system and to the vendors. The attention is focused on the data mining service provided by ASP. Those vendors who require this service and the e-commerce system pay ASP to access to DDM systems that have been provided.
- *DDM*: it is used by the ASP to provide the generic data mining systems to its subscribers. In order to support a steady working of the system in the given environment, it is necessary to own a few characteristics such as heterogeneity, infrastructure of costs, optimization, security and extensibility.

The heterogeneity involves that the software must be able to analyse the data coming from heterogeneous sources and distributed locations and to support user's requests in relation to the different calculation paradigms (client–server and mobile agents models included). The submitting philosophy is that the ASP should not impose any model on the users and that it must be able to support the specific necessities and the requests which could be provided to them.

The infrastructure of the costs refers to the system having a framework used to estimate the costs of different tasks. This implies higher calculation resources and/or a faster answer that should be more expensive on a relative scale of costs. Furthermore, the system should be able to optimize the DDM process to supply the best time of answering to the users.

The security involves that the user must manage sensitive data which should not leave the host of the site's owner. In such cases, the use of the mobile agents model has been suggested in which the data mining algorithm and the important parameters are sent to data site and where – at the end of the whole process – the mobile agent is destroyed on that same site (it never leaves the site).

The system must be extensible to supply a wide range of data mining algorithms. The user must be able to register its algorithms with ASP to use it in specific tasks of DDM. This implies that a given top-level semantic of the DDM process is necessary.

8.13 Outlines on other applications

Many DDM systems which use both client–server and mobile agents architectures have been developed: the agents' paradigm is largely used in DDM systems [17] such as PADMA (parallel data mining using agents), generic data mining, InfoSleuth, besizing knowledge through distributed heterogeneous induction (BODHI) and Papyrus.

- BODHI: it is a knowledge's discovery system based on agents, providing an environment of execution and transparent message exchange for the user.

The primary aim of BODHI project is the creation of a communication system and of a run-time environment to apply the approach of analysis of collected data without being necessarily bound to a specific platform, or to a particular learning algorithm or a given representation of knowledge.

- To avoid the limitation arising from the platforms on which the system could be used, the core of the system has been developed in Java.
- The main component is the "facilitator", a module responsible to route the data and to control the data flow among the various sites and interfaces. Each local site has a communication module, appointed as "agent station" and responsible for supplying the execution environment to the agents. The "agent stations" are furthermore responsible for the safety of the communications. The acting object inside BODHI is a Java extensible object, used as interface between the implementation of the learning algorithm and the communication module. This learning algorithm can be implemented both as extensive agent through the use of Java or as native code on the local machine. BODHI is at present under construction.
- PADMA: an approach similar to JAM, where the agents are distributed depending on the location of the various information sources, it is used in the PADMA project in which the agents execute their data mining task in parallel without melting the results.
- KEPLER: in the KEPLER system, different automatic learning algorithms have been supplemented and the concept of "extensibility" concerning data mining systems has been introduced, that is to supplement any learning algorithms inside the system. Even though it is based on the concept of "plug-in", it does not incorporate a decisional mechanism; in order to choose among the various algorithms in a given data mining session an external action is required.

8.13.1 Developing of collective data mining framework (CDM)

Kargupta *et al.* [17] developed a paradigm of distributed modelling of data and knowledge discovery called CDM. Such an approach takes its starting point from the motivation of communication theory joining them with the statistics and the automatic learning. This framework has evolved thanks to the creating of models of distributed and heterogeneous data having the guarantee of the total correctness of the model.

8.13.2 Decision trees' distributed construction coming from heterogeneous data

The authors have faced the following problem: given a set of database heterogeneous among them, a global and correct decision tree is built (compared to a decision tree built in the traditional way) using the minimum of data communication. Known techniques used to build decision trees such as the ID3 require $O(nks)$ communications where n is the number of rows of a relational table containing data, k the depth of the decision tree and s the number of distributed sites.

The local construction of the decision trees and the following efforts to aggregate the trees among them in a clever way can lead to build not-correct trees.

These approaches can furthermore not be scalable when there are too many data sources. The CDM sees the decision tree like a function to be learned through its Fourier spectrum and then converting it in a tree representation [17].

The orthonormal nature of Fourier base allows an appropriate decomposition of the tree-learning problem among different sites. Since the Fourier spectrum of a tree is scattered and it can be calculated efficiently, this approach offers the chance to generate the tree correctly in a distributed way. Once the spectrum has been estimated by the various distributed sites, it is merged in a central site and again changed to its original version of tree.

8.13.3 Global hierarchic model of clustering coming from heterogeneous data

Another problem to be faced is the following: given a collection of heterogeneous data sites, a global hierarchic model of data grouping is built for the whole set of data through the use of as less communication as possible. The global model should approximate the one which would be produced if the data were centralized.

As a possible solution, the authors suggest the generation, at each of distributed sites, of local models having a dendrogram look and the transmission of the local models to a central facilitator. It has been demonstrated that this can be achieved through an $O(n)$ communication cost. According to the limits calculated by local models, a global model approximating the "monolithic" one has been produced.

If detailed information only about a well-determined set of points is requested, the specific information concerning the values of the features can be required by the local sites.

8.13.4 Multi-varied collective regression

The multi-varied regression techniques are often used in the modelling of the data and in the knowledge discovery; therefore, the development of the distributed version of such techniques for sites having a heterogeneous nature is very interesting.

For this purpose, a CDM technique based on the wavelet for the multi-varied polynomial regression has been developed. Technically the transformation in wavelet of the data is calculated; the important coefficients from different sites are selected and collected in a central site. The regression is directly implemented on the coefficients of the wavelet and the regression model is built in the canonical representation from the model in the wavelets space.

References

[1] Botia, J.A., Garijo, J.R., and Skarmeta, A.F., A generic data mining system: basic design and implementation guidelines. *Workshop on Distributed Data Mining at the 4th Int. Conf. on Data Mining and Knowledge Discovery (KDD-98)*, AAAI Press: New York, USA, 1998.

[2] Kwon, H.C., Lee, S.I., Sohn, S., Kangl, T.G., Yoo, W.J. & Yoo, K.J., A selection algorithm for an efficient interaction pattern out of paradigms. *Workshop on Agents in Industry at the 4th Int. Conf. On Autonomous Agents*, Barcelona, Spain, 2000.

[3] Kwon, H.C. & Lee, J.T., A migration strategy of mobile agent. *Proc. of the 8th Int. Conf. on Parallel and Distributed Systems, ICPADSI*, Kyongju City, South Korea, pp. 706–712, 2001.

[4] Krishnaswamy, S., Zaslavsky, A. & Loke, S.W., An architecture to support distributed data mining services in e-commerce environments. *2nd Int. Workshop on Advanced Issues of E-Commerce and Web-Based Information Systems*, Milpitas, CA, USA, pp 239–246, 2000.

[5] Brewington, B., Gray, R., Moizumi, K., Kotz, D., Cybenko, G. & Rus, D., Mobile agents in distributed information retrieval. *Intelligent Information Agents*, ed., M. Klusch, Springer-Verlag: London, UK, pp. 355–395, 1999.

[6] Honavar, V., Miller, L., & Wong, J., Distributed knowledge networks. *Proc. of the IEEE Information Technology Conference*, Syracuse, NY, pp. 87–90, 1998.

[7] Sheth, A.P. & Larson, J.A., Federated database systems for managing distributed, heterogeneous, and autonomous databases, ACM Computing Survey, Special issue on heterogeneous databases, **22(3)**, pp. 183–236, September 1990.

[8] Wiederhold, G., The conceptual basis for mediation services. *IEEE Expert*, **12(2)**, 1995.

[9] Miller, L., Honavar, V. & Wong, J., *Object-Oriented Data Warehouses for Information Fusion from Heterogeneous Distributed Data and Knowledge Sources*, IEEE Information Technology Conference: Syracuse, NY, 1998.

[10] Glitho, R.H., Olougouna, E. & Pierre, S., Mobile Agents and their use for information retrieval: A brief overview and an elaborate case study. *IEEE Network*, Jan/Feb, **1(16)**, pp. 34–41, 2002.

[11] Stolfo, S.J., Prodromidis, A.L., Tselepis, S., Lee, W., Fan, D.W. & Chan, P.K., JAM: Java agents for meta-learning over distributed databases. *Proc. of the 3rd Int. Conf. On Knowledge Discovery and Data Mining*, AAAI Press: Newport Beach, CA, pp. 74–81, August 1997.

[12] Kong, Y.H. & Choi, I.S., An efficient web information extracting system. *Proc. of the IEEE Int. Symposium on Industrial Electronics*, ISIE 2001, 1–16 June 2001, Pusan, South Korea, **3**, pp. 1771–1774.

[13] http://www.gsi.dit.upm.es/~mast/

[14] King, R.D., Feng, C. & Shutherland, A., STATLOG: comparison of classification algorithms on large real-world problems. *Applied Artificial Intelligence*, **9(3)**, pp. 259–287, May/June 1995.

[15] Yang, J., Honavar, V., Miller, L. & Wong, J., Intelligent mobile agents for information retrieval and knowledge discovery from distributed data and knowledge sources. *IEEE Information Technology Conference*, Syracuse, NY, pp. 99–102, September 1998.

[16] Yang, J., Pai, P., Honavar, V. & Miller, L., Mobile intelligent agents for document classification and retrieval: A machine learning approach. *Proc. of the 14th European Meeting on Cybernetics and Systems Research*. Symposium on Agent Theory to Agent Implementation, Vienna, Austria, 1998.

[17] Kargupta, H., Distributed Knowledge Discovery from Heterogeneous Sites, available at http://www.cs.umbc.edu/~hillol/DKD/ddm_research.html

Index

WITPRESS ...*for scientists by scientists*

Risk Analysis IV

Edited by: ***C.A. BREBBIA****, Wessex Institute of Technology, UK*

This book contains over 70 papers from the fourth in this popular international conference series. Topics covered include: Seismic Risk; Floods and Droughts; Man-Made Risk; Estimation of Risk; Risk Assessment and Management and Risk Mitigation.

Contributions from three special sessions highlighting the work of renowned international experts are also featured. These deal with Geomorphic Hazard and Risk, Seismic Risk Analysis in Mediterranean Cities and Landslides from Hazard to Risk Prevention.

WIT Transactions on Ecology and the Environment, Vol 77

ISBN: 1-85312-736-1 2004 832pp £291.00/US$465.00/€436.50

Text Mining and its Applications to Intelligence, CRM and Knowledge Management

Edited by: ***A. ZANASI****, TEMIS Text Mining Solutions SA, Italy*

Organizations generate and collect large volumes of textual data. Unfortunately, many companies are unable to capitalize fully on the value of this data because information implicit within it is not easy to discern.

Primarily intended for business analysts and statisticians across multiple industries, this book provides an introduction to the types of problems encountered and current available text mining solutions.

Partial Contents: Text Processing and Information Retrieval; Application Integration in Applied Text Mining; ROI in Text Mining Projects; Open Sources Automatic Analysis for Corporate and Government Intelligence; Customer Feedbacks and Opinion Surveys Analysis in the Automotive Industry; Text Mining Based Knowledge Management in Banking; Text Mining in Life Sciences and Link Analysis in Crime Pattern Detection.

Series: Advances in Management Information Vol 2

ISBN: 1-85312-995-X 2005 368pp+CD-ROM £144.00/US$230.00/€216.00